AF798719

Digital Innovations in Architecture, Engineering and Construction

Series Editors

Diogo Ribeiro, Department of Civil Engineering, Polytechnic Institute of Porto, Porto, Portugal

M. Z. Naser, Glenn Department of Civil Engineering, Clemson University, Clemson, SC, USA

Rudi Stouffs, Department of Architecture, National University of Singapore, Singapore, Singapore

Marzia Bolpagni, Northumbria University, Newcastle-upon-Tyne, UK

The Architecture, Engineering and Construction (AEC) industry is experiencing an unprecedented transformation from conventional labor-intensive activities to automation using innovative digital technologies and processes. This new paradigm also requires systemic changes focused on social, economic and sustainability aspects. Within the scope of Industry 4.0, digital technologies are a key factor in interconnecting information between the physical built environment and the digital virtual ecosystem. The most advanced virtual ecosystems allow to simulate the built to enable a real-time data-driven decision-making. This Book Series promotes and expedites the dissemination of recent research, advances, and applications in the field of digital innovations in the AEC industry. Topics of interest include but are not limited to:

- Industrialization: digital fabrication, modularization, cobotics, lean.
- Material innovations: bio-inspired, nano and recycled materials.
- Reality capture: computer vision, photogrammetry, laser scanning, drones.
- Extended reality: augmented, virtual and mixed reality.
- Sustainability and circular building economy.
- Interoperability: building/city information modeling.
- Interactive and adaptive architecture.
- Computational design: data-driven, generative and performance-based design.
- Simulation and analysis: digital twins, virtual cities.
- Data analytics: artificial intelligence, machine/deep learning.
- Health and safety: mobile and wearable devices, QR codes, RFID.
- Big data: GIS, IoT, sensors, cloud computing.
- Smart transactions, cybersecurity, gamification, blockchain.
- Quality and project management, business models, legal prospective.
- Risk and disaster management.

Erika Dolníková

Simulation Tools for Predicting Daylighting Conditions in Buildings

Springer

Erika Dolníková
Faculty of Civil Engineering
Institute of Architectural Engineering
Technical University of Kosice
Košice, Slovakia

ISSN 2731-7269 ISSN 2731-7277 (electronic)
Digital Innovations in Architecture, Engineering and Construction
ISBN 978-3-031-78329-6 ISBN 978-3-031-78330-2 (eBook)
https://doi.org/10.1007/978-3-031-78330-2

© The Editor(s) (if applicable) and The Author(s), under exclusive license to Springer Nature Switzerland AG 2025

This work is subject to copyright. All rights are solely and exclusively licensed by the Publisher, whether the whole or part of the material is concerned, specifically the rights of translation, reprinting, reuse of illustrations, recitation, broadcasting, reproduction on microfilms or in any other physical way, and transmission or information storage and retrieval, electronic adaptation, computer software, or by similar or dissimilar methodology now known or hereafter developed.
The use of general descriptive names, registered names, trademarks, service marks, etc. in this publication does not imply, even in the absence of a specific statement, that such names are exempt from the relevant protective laws and regulations and therefore free for general use.
The publisher, the authors and the editors are safe to assume that the advice and information in this book are believed to be true and accurate at the date of publication. Neither the publisher nor the authors or the editors give a warranty, expressed or implied, with respect to the material contained herein or for any errors or omissions that may have been made. The publisher remains neutral with regard to jurisdictional claims in published maps and institutional affiliations.

This Springer imprint is published by the registered company Springer Nature Switzerland AG
The registered company address is: Gewerbestrasse 11, 6330 Cham, Switzerland

If disposing of this product, please recycle the paper.

About This Book

This book explores the significant role of natural daylight in influencing physical and mental health, with a focus on sleep, mood, anxiety, stress, and productivity. Surveys reveal that exposure to daylight has notable effects on well-being, particularly in countries like Italy and Slovakia, where it impacts mood and productivity. The importance of outdoor activities and adequate exposure to morning daylight is emphasized, as it helps regulate the circadian rhythm and improves sleep quality. Modern society, however, spends most time indoors, where lighting often falls short of stimulating the circadian system, affecting health and productivity. The chapter also addresses workplace lighting, highlighting its impact on visual fatigue, well-being, and safety. Using simulation tools like RADIANCE and Velux Daylight Visualizer, it evaluates light comfort in different building types, offering insights into the future impact of lighting on energy efficiency and user comfort. The chapter is structured into four sections: an overview of current issues, computer simulation techniques, daylight measurement in buildings, and case studies on predictive lighting modeling.

Contents

Symbols and Acronyms

Capital Letters

DF_e	External reflected component of the daylight factor (%)
DF_i	Internal reflected component of the daylight factor (%)
DF_{max}	Maximum value of the daylight factor (%)
DF_{min}	Minimum value of the daylight factor (%)
DF_{avg}	Average value of the daylight factor (%)
DF_S	Sky component of the daylight factor (%)
E_{eH}	Horizontal external illuminance (lx)
E_{ie}	External reflected illuminance (lx)
E_{ir}	Internal reflected illuminance (lx)
E_{is}	Interior illuminance (lx)
E_v	Vertical illuminance (lx)
G	Glare index (–)
GI	Glare Index for artificial lighting (–)
K_p	Position factor of the glare source in the field of view (–)
L	Brightness of the radiated surface (cd m^{-2})
L_e	Radiant glow (W m^{-2} sr^{-1})
L_i	Falling glow (W m^{-2} sr^{-1})
L_p	Background brightness, or adaptive brightness (cd m^{-2})
L_r	Reflected glow (W m^{-2} sr^{-1})
L_S	Brightness of the source (cd m^{-2})
L_Z	Brightness of the sky at the zenith (cd m^{-2})
L_ε	The brightness of the sky in the angle ε and horizontom (cd m^{-2})
R	Total exposure (W m^{-2} sr^{-1})
S	Glare factor (–)

Small Letters

t_Z Glazing pollution factor (–)

Greek Letters

ω Spatial angle in which the dazzling source (sr)
ε Elevation angle (°)
ρ Average reflection coefficient of all internal surfaces (–)

Acronyms

ČSN	Czechoslovak standard
CIE	International Commission on Illumination
DF	Daylight factor (in graphic dependencies)
DGI	Daylight Guth Index
DGP	Daylight Guth Probability
DGPs	Simplied Daylight Guth Probability
IESNA	Illumination Engineering Society
STN	Slovak technical standard
UGP	Unified Glare Probability
UGR	Unified Glare Rating
UV	Ultraviolet (radiation)
VCP	Visual Comfort Probability

List of Figures

List of Tables

Part I
Simulation Tools for Predicting Daylighting Conditions in Buildings

Introduction

In our lives, natural daylight plays a very significant role. Daylight has a positive impact on our physical and mental health. Studies on the effects of vitamin D and UV radiation on the environment have raised general awareness of natural light. Exposure to indoor daylight and time spent outdoors have a substantial impact on all survey outcomes, including sleep disorders, anxiety, stress, depression, and mood.

According to survey findings, up to 47% of Slovaks stated that daylight affects their mood. A survey conducted by one company revealed that more than half of people (52 %) believe that daylight affects their sleep. Two-thirds of respondents (63 %) think that daylight influences their productivity. The surveys also indicated that daylight has the most impact on residents of Italy and Slovakia, while it has the least impact on residents of the United Kingdom.

Another important factor is the amount of time spent in outdoor environments. The survey results revealed that people who spent one to two hours outside each day experienced significantly less anxiety, stress, and depression, and their sleep was better compared to those who spent less than 30 minutes outdoors each day. Moreover, morning daylight provides the greatest benefits. The results show that even just one day spent outdoors can help improve mood and contribute to better sleep at night.

In contemporary society, people typically spend more than 90 %of their time indoors, but the lighting in built environments, particularly in homes, is usually not sufficient to stimulate the circadian rhythm.

Typical indoor lighting provides less than 100 lx, while being outdoors on a sunny day can provide anywhere from 1, 000 lx to 3, 000 lx, and up to 100 klx, on a bright sunny day. Exposure to a 24-hour light and dark cycle supports the circadian rhythm, which has many health benefits, such as increased alertness and feelings of vitality during the day.

Lighting is one of the main conditions of the work environment. Properly designed lighting conditions for work prevent visual fatigue and enhance workplace well-being, productivity, and accuracy of work performed. Insufficient lighting in work-places can negatively affect productivity and work quality. Therefore, highquality lighting is key to safe operations, increasing work capacity, and reducing the risk of workplace injuries.

The presented chapter focuses on the issue of evaluating light comfort in selected buildings such as production halls, administrative buildings, and school buildings. The assessment and prediction of light are carried out using the simulation program RADIANCE and the Velux Daylight Visualizer program. Modeling and computer simulations allow the prediction of future impacts of light comfort not only from the perspective of energy efficiency but also in terms of user comfort in relation to the entire building.

The chapter is divided into four comprehensive sections. The aim of the first part is to summarize the current issues in our country and abroad. The second part is dedicated to computer simulations and specifically to the RADIANCE program, which was used to evaluate the selected objects. The third part focuses on describing the measurement of daylight in buildings. The final part highlights the possibilities of modeling and predicting the lighting environment of buildings through computer simulations in case studies.

This paper was elaborated with the financial support of the research project VEGA1/0626/22 of the Scientific GrantAgency, the Ministry of Education, Science, Research, and Sport of the Slovak Republic and the Slovak Academy of Sciences.

Chapter 1
Current State of the Issue

Over the course of several centuries, there has been a revolutionary change in the quality, quantity, and temporal availability of light. The regular rhythm of alternating day and night has disappeared over the past 200 years. Today, we live in an environment where daylight levels during the day are far below the necessary values, while after sunset we are regularly exposed to higher light intensities than ever before.

After a period around the mid-20th century, during which daylight was not considered essential due to cheap energy, abundant raw materials and fuels, and a lack of foresight about future development, its importance was revived in the 1970s as a result of the oil crisis and later environmental concerns. Natural light is the most valuable source of interior lighting for building inhabitants.

The share of electricity consumption for lighting in commercial buildings in Slovakia ranges from 35 to 50%, while in industry it is about 10%. The penetration of natural light into buildings and the influence of light and thermal components on energy consumption can lead to significant cost reductions.

However, the use of natural light does not come without challenges. Issues with glare, overheating, variability, and privacy can cause problems. Therefore, building design must regulate the penetration of natural light into the building, reducing energy consumption and creating a visually stimulating environment. Daylight is a fundamental human need, emphasized by providing transparent and translucent surfaces in all inhabited spaces we see around us, and is a major factor in properly setting the circadian rhythm.

1.1 Work Environment and Lighting Comfort

Currently, there are numerous studies that examine daylight and its impact on the indoor environment. The work environment is created by the workspace, which in industrial settings includes machines and equipment, work tools, transportation

© The Author(s), under exclusive license to Springer Nature Switzerland AG 2025

E. Dolníková, *Simulation Tools for Predicting Daylighting Conditions in Buildings*,
Digital Innovations in Architecture, Engineering and Construction,
https://doi.org/10.1007/978-3-031-78330-2_1

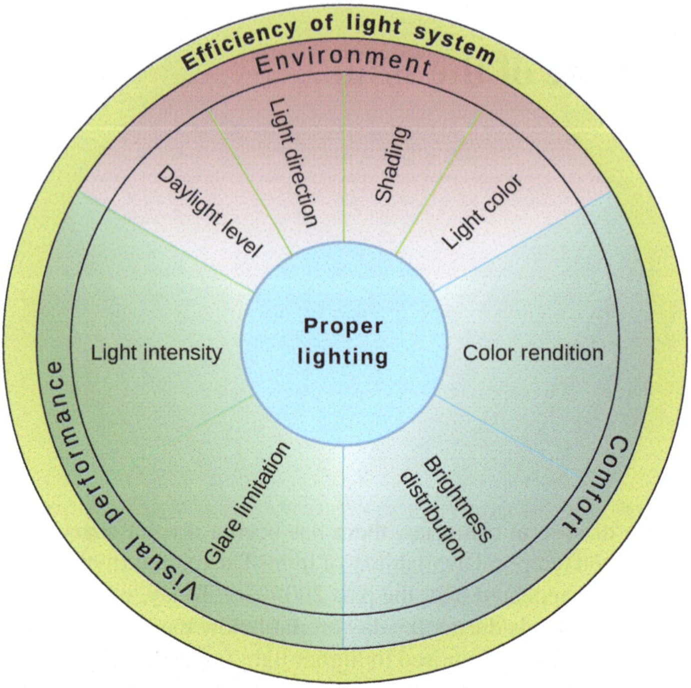

Fig. 1.1 Proper lighting [3]

means, lighting, thermal and acoustic conditions, color schemes, and also the social climate. The quality of lighting in the workplace is crucial for both employee well-being and productivity. Proper lighting can reduce eye strain, improve mood, and enhance overall performance, while poor lighting can lead to discomfort, fatigue, and even health problems. Significant importance is placed on the strain on the human body from various types of work activities, the balance of work and rest, hygiene, and work organization. In a narrower sense, creating a work environment involves a combination of factors that influence the overall well-being of employees in the workplace, see Fig. 1.1, [1–4].

Lighting is an important element of the work environment that affects the quality, efficiency, and safety of work. Natural daylight is indispensable for humans and should be used purposefully and efficiently. Daylight varies greatly throughout the day, with changes in the amount, intensity, and spectral composition of light. It also depends on the season. Unrestricted access to natural light may not always be the best for visual comfort. Therefore, the amount of light in workplaces needs to be regulated. Fluctuations in daylight can be compensated with electric lighting to achieve the desired amount and uniformity of illumination indoors [5–7].

Due to the ongoing energy crisis in many countries, using natural light in buildings has become a necessity rather than just a preference. Besides reducing energy consumption for lighting, natural light is reported to boost occupants' moods and increase productivity in workplaces. Among the various daylighting systems available today, top lighting is the most commonly used due to its ability to provide daylight with relatively homogeneous distribution, especially in deep rooms where side lighting cannot adequately reach the back of the space. Additionally, the wide range of top lighting strategies allows the system to be used for both aesthetic and energy purposes. The usefulness of top lighting is expected to increase with the current rate of urbanization [5, 7, 8].

1.2 Daylighting

The sun is an inexhaustible source of energy that influences the environment and is the primary source of light and heat. Some types of electromagnetic waves cannot be directly perceived or felt by humans. Humans can immediately perceive and feel only a relatively narrow range of the electromagnetic spectrum. According to the wavelength, solar radiation is divided into the following categories [4, 9, 10]:

- Ultraviolet Radiation (UV) with wavelengths $\lambda = 100–380$ nm,
- Visible Light with wavelengths $\lambda = 380–780$ nm,
- Infrared Radiation (IR) with wavelengths $\lambda = 780–10^6$ nm.

Light, as the most important part of the solar radiation spectrum, is indispensable for human existence. Solar radiation emitted by the Sun undergoes attenuation and changes its original character as it passes through the Earth's atmosphere. Therefore, at the Earth's surface, when determining the amount of solar radiation received, there are three components of solar radiation apart from the horizontal plane [11, 12].

It consists of direct radiation, which is directed straight (more or less unchanged in direction) from the solar disk, further diffuse radiation, scattered by atmospheric particles and coming from all directions of the celestial hemisphere, and finally, radiation reflected from surrounding terrain. These three components of solar radiation together constitute total solar radiation, unlike global solar radiation, which strikes the horizontal plane, containing only direct and celestial diffuse radiation components [4, 12].

Skylight is formed by scattering, reflection, and transmission of sunlight through the atmosphere, through its gaseous and particulate components. Its effect on the Earth's surface depends on the sun's position, atmospheric conditions, and cloud cover. Lighting design and assessment are based on the most adverse conditions of winter daylight regimes—completely and evenly overcast skies. A characteristic feature of daylight is its variability, which concerns illuminance, the direction of direct sunlight versus skylight, and spectral composition. Daylight not only conditions our vision but also significantly influences human health [4, 13–15].

Appropriate lighting, for instance in manufacturing facilities, becomes a crucial factor in production rationalization, enabling increased work productivity, reduced incidence of defective products, and thereby improving production quality. Lighting comfort is influenced by various factors, such as [4, 7, 16, 17]:

- illuminance intensity (in lux),
- uniformity of lighting,
- method of lighting shading,
- elimination of glare,
- light color.

Indoor spaces can be illuminated as follows:

1. daylighting (natural light);
2. artificial lighting (electric light sources);
3. combined lighting (daylight and artificial light).

The fundamental quantitative characteristic of daylighting is illuminance. The primary qualitative characteristics of daylighting include the uniformity of lighting, distribution of luminous flux, distribution of surface brightness in the field of view, occurrence of glare, and the color scheme of the interior. Daylighting in interiors is evaluated quantitatively by the daylight factor. The daylight factor characterizes the lighting engineering properties of indoor spaces within a building, depending on the design of the lighting system [4, 7, 18, 19].

It is expressed as the ratio of the daylight illuminance penetrating through window constructions onto the evaluated workplace to the maximum exterior illuminance on an unshaded horizontal plane, under conditions of graduated luminance distribution on a completely overcast sky.

The daylight factor is determined by calculation or measurement in a completed building, or by measurement on a scale model. The distribution of luminance on a uniformly overcast sky over a dark terrain is calculated according to the relationship [20]:

$$L_{\varepsilon} = \frac{1}{3} L_{Z} \left(1 + 2 \sin \varepsilon\right), \tag{1.1}$$

The distribution of luminance on a uniformly overcast sky over a snowy terrain is calculated according to the formula [20]:

$$L_{\varepsilon} = \frac{1}{2} L_{Z} \left(1 + \sin \varepsilon\right), \tag{1.2}$$

where L_{ε} je brightness of the sky at an angle ε above the horizon (cd/m^2),

L_Z—brightness of the sky at the zenith (cd/m^2).

The access of daylight into indoor spaces should be oriented to maintain an uninterrupted view through lighting openings towards the surroundings. This is primarily for psychological reasons, to ensure necessary visual connection with the external environment. It is also important to ensure that daylight does not cause glare from the sun or sky, as this could adversely affect thermal comfort [4, 21, 22].

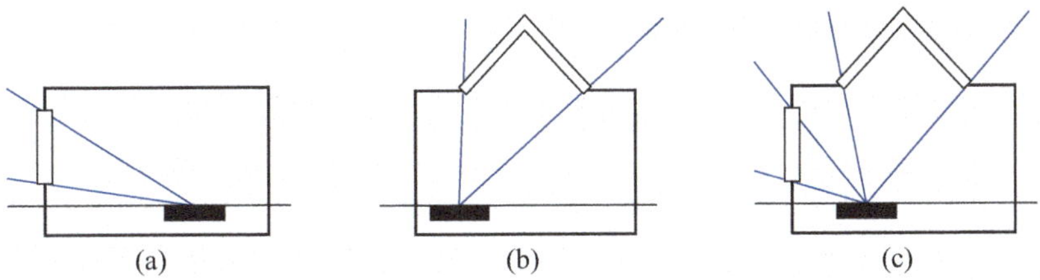

Fig. 1.2 Basic methods of daylighting: **a**—side, **b**—overhead, **c**—combined [23]

1.3 Systems of Building Illumination

According to the placement of lighting spaces in the building envelope, lighting systems are see Fig. 1.2, [23, 24]:

1. side,
2. overhead,
3. combined.

1.3.1 Side Lighting System

The daylighting system is implemented through side lighting openings (windows) in the building envelope, particularly for multi-story buildings. This system is characterized by uneven illumination of the horizontal work plane with vertical daylight sources. Only areas close to the windows are effectively illuminated [22].

The shape, size (window to wall ratio—WWR), and types of glazing should be considered as part of building lighting systems in workplaces to balance the requirements of daylighting against the need to reduce solar gains [25, 26].

According to many studies, the primary advantage of windows is their provision of external views, followed by access to daylight. Optimal window design requires adherence to principles such as: window size (determined by WWR), type of glazing, comparison of different glazing types with varying U-values and transmittance, and consideration that electrochromic glazing, while providing good illumination, may not always be energy-efficient. In offices, the window area should ideally constitute at least 20–25% of the wall area, preferably 30–25% and more [25–28].

1.3.2 Top Lighting System

This system is the predominant method for utilizing daylight in industrial interiors. In a top lighting system, overhead lighting openings uniformly distributed in the ceiling or roof enable achieving uniform daylighting without significant variations.

Therefore, it is advantageous to illuminate the horizontal work plane with horizontal daylight sources (various types of skylights). Hence, the top lighting system is more effective than the side lighting system in terms of daylighting [4, 29, 30].

It is also important to ensure the exclusion of direct sunlight and unwanted glare. Typically, top lighting systems are used to illuminate extensive areas of ground-level industrial halls, especially in central aisles. Apart from multi-sided and zenith top lighting systems, other lighting openings are elongated and have a 'shape'. Depending on their orientation relative to the longitudinal axis of the room's floor plan, these are referred to as transverse or longitudinal skylights.

Some types of skylights are more suitable for so-called single-aisle halls. The appropriate selection in a specific case depends on the comprehensive solution of operations, technological requirements, structural span requirements, ventilation needs, and other production requirements, as well as conditions for creating an optimal working environment [4, 31–33].

Most double-sided top lighting systems are designed either for single-aisle assembly halls or with multiple repetitions for manufacturing interiors where larger spatial objects are handled. Compared to single-sided skylights, double-sided skylights are used for larger spans. Despite having double the glazed area, they do not achieve as high a level of daylighting due to their design.

For industrial buildings, the predominant choice is often strips of ridge skylights, either transverse or longitudinal. With a zenith lighting system and appropriate even distribution of these skylights, it is possible to achieve uniformly distributed lighting over horizontal surfaces overall [24, 31, 34, 35].

Sawtooth skylights represent the best choice for providing daylight in industrial buildings [35]. Study results indicate that skylight openings ranging from 3 to 14% of the ceiling area can reduce energy consumption by up to 19% [7, 36]. Among top lighting systems, light pipes are included (there exists the HOLIGILM program), which can calculate the lighting of rectangular rooms using optical tubes as light sources [37, 38].

1.3.3 Combined Lighting System

It's a combination of side and top lighting systems within a single indoor space. The effectiveness of combined lighting can be seen in the Fig. 1.3 [23].

Their main feature is the relative difficulty in predicting adequate daylight illumination. It is used in single-story buildings or on the top floor of multi-story buildings, where top lighting eliminates the unevenness of the side lighting system. While

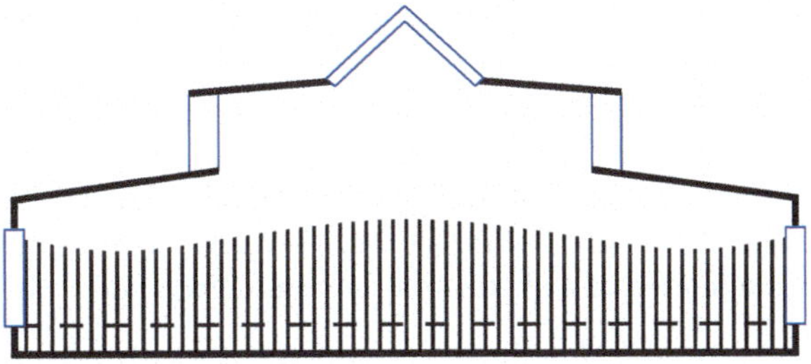

Fig. 1.3 The effectiveness of combined lighting [23]

skylights distribute daylight, window constructions allow views of the outside environment. This system often provides the only suitable solution in certain challenging situations, not only in the design of industrial buildings [31, 39].

In buildings with a top lighting system, side window bands are installed to allow views and contact with the exterior. The combined lighting system ensures better light distribution in industrial operations [24, 26, 31, 32, 35, 40].

The amount of light that skylights let in depends on the amount of daylight available from the external environment, which varies with climatic conditions throughout the day and over the year. On clear sunny days, the maximum amount of daylight is available. On very dark rainy days, there is less light. Skylights have been used in the architectural and construction industry for some time, and scientists have quantified and confirmed their potential for building sustainability.

Many authors have researched open skylights, their advantages, and disadvantages. They concluded that lighting through skylights is more efficient than through windows, both in terms of light levels and uniformity. Using the flow transfer method developed by Bouchet and Fontoynont, the authors studied the efficiency of circular skylights and the amount of light these skylights transmitted into a room [5, 35, 41–44].

Conventional windows effectively meet basic daylighting needs in spaces, but advancements in technology offer solutions that surpass traditional capabilities. Daylighting systems play a crucial role in optimizing daylight performance and are integral to a comprehensive daylighting strategy. Careful selection of these systems is essential as inappropriate choices can undermine the potential benefits of well-designed spaces. Conversely, a well-chosen daylighting system can compensate for early design-stage [45].

Moreover, daylighting systems contribute significantly to reducing both thermal discomfort and visual glare caused by direct sunlight through controlled and uniform distribution of daylight onto work planes. This not only enhances daylighting quality and performance but also improves overall visual comfort for occupants. Furthermore, such systems facilitate increased utilization of daylight in sunny climates where effective daylight control is critical. To select the most suitable daylighting system, several key parameters must be carefully considered as initial steps [45]:

1. factors such as building site characteristics (including latitude, cloud cover, and obstructions) and prevailing daylight conditions,
2. daylighting strategies integrated into the architectural design process,
3. design and functional aspects of windows,
4. operational constraints of the system, including fixed or operable components,
5. functional capabilities and performance metrics of the system,
6. architectural and construction limitations and considerations.

It is crucial to highlight the primary objectives behind implementing daylighting systems [45]:

1. redirecting daylight effectively,
2. enhancing daylight quality to improve task illumination and control glare,
3. providing solar shading to enhance thermal and visual comfort.

1.4 Advantages of Daylighting

With a sufficient level of daylight, a good daylighting design allows for the intensity of artificial lighting to be reduced or even turned off completely. This situation helps reduce the building's electricity consumption during peak demand periods, which lowers the overall energy cost as measured in kWh [46–48].

In terms of luminous efficacy, which expresses the amount of lumens per watt of thermal energy, the efficiency of daylight is higher than that of commonly used artificial lighting methods. Since daylight produces less heat per unit of illumination than artificial lighting and almost all the energy from the light source is converted into heat within the building, using daylighting can reduce cooling and heating costs.

The overall goal of daylighting is to minimize the amount of artificial light used and reduce electricity costs, but it can also lower costs related to thermal comfort and acceptable indoor air quality. Electric lighting produces a lot of heat, whereas properly controlled natural lighting produces almost no heat [35, 49, 50].

In most buildings with daylighting, overall energy savings range from 15 to 40%. While energy savings and sustainability might be the initial reasons companies choose daylighting, it can also impact the productivity and satisfaction of employees, students, clients, and retail customers [7, 49, 51, 52].

People naturally need daylight. Studies suggest that daylight has a direct impact on well-being, productivity, and overall satisfaction. Even retail stores see environmental and financial benefits from daylighting for both employees and consumers. Numerous experiments have shown that stores with skylights over certain departments had higher overall sales per square meter compared to departments illuminated by artificial light [49, 53].

It has also been found that students with more daylight in their classrooms progressed 20% faster in math tests and 26% faster in reading tests over the course of a year [49, 54–56].

For daylighting to be cost-effective, the savings on lighting and cooling must offset the costs of purchasing, installing, and operating the daylighting system within a reasonable timeframe.

The amount of energy savings depends on climatic conditions, location, energy load, and building design. Savings or gains in the workspace are difficult to quantify. Therefore, the impact of these changes on energy costs is not negligible [36, 52, 57].

1.5 Disadvantages of Daylighting

One significant disadvantage of daylighting is the inability to precisely determine its intensity in advance. Factors influencing adequate daylight access include the sun's position and intensity, cloud cover, and shading from nearby terrain features, vegetation, and objects [4, 35, 54].

Improperly designed daylighting can lead to glazed areas through which daylight enters the building contributing to heat loss in winter and unwanted heat gain in summer. The costs of additional heating and cooling then offset savings from reduced lighting levels. Insufficient lighting can cause accidents and errors, waste energy in lighting, and strain the eyes with improperly positioned light fixtures [29, 35, 58].

Direct sunlight in classrooms and office spaces often causes unpleasant glare on work surfaces, making it difficult to work or view computer screens. Significant amounts of heat or light are not the only challenges associated with daylighting strategies. Some architectural elements, such as the building's roof, atrium shapes, or building angles, can prevent daylight from effectively illuminating a space.

1.6 Visual Comfort and Glare

Visual comfort is defined as the state of mind expressing satisfaction with the visual environment. Adequate levels of light are crucial; insufficient or excessive illumination can lead to discomfort among occupants, impacting their satisfaction and productivity. Factors influencing visual comfort include light distribution, light quality, glare, and overall lighting levels, which are critical to meeting human needs. Glare, in particular, has emerged as a key criterion for evaluating visual comfort, defined as "the sensation" caused by luminance within the visual field significantly exceeding the luminance to which the eyes are accustomed, resulting in annoyance, discomfort, or diminished visual performance and visibility [59–62].

According to the International Commission on Illumination (CIE), glare is defined as a condition where there is discomfort glare or reduction in the ability to see significant objects due to inappropriate distribution or range of luminances, extreme contrasts in space or time. In the study of discomfort glare, the most common method currently used involves subjective evaluation of glare on glare scales, and glare index values are calculated from measured physical quantities. There are four main physical quantities on which discomfort glare indices are based [59]:

- the luminance of the source(s) causing glare in the field of view,
- the angle of the source(s) causing glare in the field of view,
- the luminance of the background,
- the position index.

Effective measurement of these quantities in situ employs a technique using high dynamic range (HDR) imaging. This technique is used to create luminance maps containing the subject's field of view. Illuminance indices are then computed based on these luminance maps. Alternatively, these luminance maps can also be generated using simulation tools.

Calculating the illuminance index based on the luminance map is therefore a necessary step and an integral part of studying uncomfortable glare. However, the difficulty lies in identifying the source(s) of glare in the field of view. Once a glare source is identified on the luminance map, its angle, position index, and brightness can easily determine the illuminance index.

Background brightness can be quickly evaluated as the average brightness from the luminance map, excluding the glare source. However, defining the glare source remains ambiguous, and there is no universal method for its identification [59].

Visual comfort depends on the quantity and quality of lighting, spatial characteristics, and the state of the visual organ. Visual work is also influenced by the relationship between lighting (its quantity), object size, and the time allotted for visual tasks. The finer the detail and the shorter the time, the greater the demands on lighting [59].

A sufficient amount of daylight alone does not ensure visual comfort. The quality of daylight depends mainly on the distribution of luminous flux, the predominant direction of lighting, the distribution of brightness on main surfaces in the observer's field of view, and the occurrence of undesired phenomena, especially glare. Besides visual comfort, another criterion characterizing the level of vision is visual performance, determined by physiological characteristics of human vision.

The issue of lighting conditions can be assessed from two fundamental perspectives: the characteristics and principles of human vision at work and the principles that must be respected when creating conditions for vision, i.e., principles for various types of lighting [4, 59, 63, 64].

Assessing visual comfort is a crucial aspect of architecture and building design as it directly impacts the comfort and productivity of users. Visual comfort encompasses several factors, including the quality of lighting, illumination, glare, colors, and contrasts. Various methods and standards are used worldwide to evaluate visual comfort. Different countries and international organizations have established norms and standards for visual comfort, focusing on lighting and other visual aspects. Some of the most important ones include [65, 66]:

- EN 17037 (Europe): European standard for daylight in buildings, specifying requirements for lighting in rooms to ensure sufficient daylight, reduce glare, and promote visual comfort,

- EN 12464-1 (Europe): European standard specifying lighting requirements for indoor workplaces,
- IES LM-83 (USA): Standard by the Illuminating Engineering Society, detailing methods for evaluating daylight, including Annual Sunlight Exposure (ASE) and Spatial Daylight Autonomy (sDA),
- ISO 8995-1: International standard for workplace lighting.

Various quantitative indicators are used to measure lighting quality and other relevant factors [65, 66]:

- Illuminance: The amount of light falling on a surface, measured in lux (lx). Optimal values depend on the type of space and activity performed,
- Uniformity: The ratio between minimum and average illuminance in a space, ensuring even light distribution,
- Glare: Unwanted light that reduces visual comfort, assessed using indices like Unified Glare Rating (UGR) or Daylight Glare Probability (DGP),
- Daylight Factor (DF): The ratio between indoor and outdoor illumination levels, expressed as a percentage. Used to evaluate the amount of daylight in a room,
- Spatial Daylight Autonomy (sDA): The percentage of floor area that meets the minimum required illuminance (e.g., 300 lux) for a specified portion of the work-day and year,
- Annual Sunlight Exposure (ASE): The percentage of floor area exposed to high levels of direct sunlight annually, potentially leading to glare or overheating.

Visual comfort is also a subjective experience, assessed through:

- Questionnaires and Surveys: Collecting user feedback through questionnaires that evaluate satisfaction with lighting and visual comfort,
- Qualitative Observations: Direct observations and interviews with users, providing deeper insights into their experiences and needs.

Light quality includes factors such as:

- Spectral Quality: The composition of the light spectrum, influencing color perception and visual comfort.
- Color Temperature (CCT - Correlated Color Temperature): Measured in Kelvins (K), affecting the atmosphere and comfort in a space,
- Color Rendering Index (CRI): The ability of a light source to accurately reproduce colors, measured on a scale from 0 to 100.

Certification systems and sustainability assessments often include criteria for visual comfort:

- LEED (Leadership in Energy and Environmental Design): Certification system evaluating indoor environmental quality, including lighting,
- WELL Building Standard: Certification focusing on health and well-being of users, emphasizing lighting and visual comfort.

Assessing visual comfort involves a comprehensive approach that combines quantitative measurements, simulations, subjective evaluations, and adherence to standards and norms. These methods and tools ensure that the indoor environment of buildings is designed to provide optimal visual conditions for users, enhancing their comfort and productivity while promoting energy efficiency and sustainability.

1.6.1 Glare

Glare is a visual discomfort. It can have various dimensions [66]:

- access to daylight,
- readability on screens,
- external view,
- contrast in the field of view,
- color,
- glare.

Glare is always caused by a large-area source (window, skylight, sunlit surface, etc.) that has relatively high luminance. Skylights are not usually designed for contact with the external environment. Therefore, daylighting systems such as light tubes or non-transparent systems can be used in their design. Regulating glare with skylights is much simpler than with side lighting, which must provide users with contact with the external environment [4, 32, 59].

In top lighting, the issue of solar shading is important. One method for designing top lighting in sunny climates involves using innovative daylighting systems to achieve a homogeneous distribution of light in the space.

Shading devices for skylights—such as sun protection, prismatic panels, and directional selective shading systems using holographic optical components—are applied to extensive glazed roof areas in higher latitudes. When situated on a window pane, they are protected from dust and require less demanding maintenance. These systems require an individual approach for application [4, 35, 40].

The disadvantage of skylights compared to windows is that they gather more light and heat in the summer than in the winter—typically, the opposite is desired. For this reason, vertical or nearly vertical skylights—such as clerestory windows, sawtooth, or monitor skylights—are often implemented to light single-story spaces. Depending on the circumstances, they can be oriented to the north, south, east, or west and shaded with conventional devices.

The consistency and quality of daylight can be enhanced by placing light-capturing louvers on the external glazing and distributing the reflected light into the space depending on the geometry of the roof skylight [46, 55].

The glare occurs when a disproportionately high brightness acts on the retina or a part of it relative to the adaptation brightness. Glare can be [4]:

absolute glare from direct sunlight,
glare by contrast glare where surfaces with different brightness levels are present in the field of view,
transitional occurs with sudden changes in brightness in the field of view,
veiling caused by reflections from mirrored surfaces.

A higher degree of glare is *restrictive* glare, which makes recognition difficult and evokes feelings of visual strain, tension, uncertainty, and fatigue, and *blinding* glare, which makes seeing impossible.

Glare can occur not only from direct exposure to a dazzling source but also from its reflection (e.g., sunlight reflected in a window opposite a stationary building). Therefore, based on the positioning of the dazzling source in the field of view, we distinguish between *axial* (foveal) and *eccentric* (peripheral) glare. This means that the brightness that is intolerable in axial glare may only be unpleasant in peripheral glare. This depends on the angular deviation from the axis of the field of view [4, 59].

1.6.2 Assessment and Determination of the Degree of Glare

The magnitude and nature of glare are expressed by the glare factor. It is quantified using several empirical relationships. In our context, the most practical is the Netusil formula [4]:

$$S = \frac{K_p\, L_S\, \omega^{0.4}}{\sqrt{L_p}}, \tag{1.3}$$

where K_p is position index of the dazzling source in the field of view,
L_S – luminance of source (cd/m^2),
L_p – background luminance, or adaptation luminance (cd/m^2),
ω – solid angle of source (sr).

In the STN 73 0580-1 standard, it is stated that the ratio of the luminance of the observed object to the luminance of the sky in the illuminating aperture behind it should not exceed 1:200. However, currently, standardized conditions and a normative method for calculating luminance ratios in interiors are lacking.

To achieve comfortable visual conditions, the ratio of the luminance of the observed object to the average luminance of the surfaces immediately surrounding it should be within 3:1 (or 1:3), and for distant surfaces, it should be 10:1.

The discomfort glare caused by viewing the sky through a window can be determined by the 'glare index', derived from the Cornell equation for extended glare sources.

$$G = \frac{K\, L_S\, \Omega^{0.8}}{L_b + 0.07\, \omega^{0.5} L_S}, \tag{1.4}$$

where K is constant dependent on the units used,

L_S – luminance of source (cd/m^2),
L_b – background luminance (cd/m^2),
Ω – solid angle of source (–).

The angle Ω can be determined using the method developed by Petherbridge and Longmore. The glare source must be depicted on the diagram, and its area is expressed as a fraction of the total area.

$$GI = 10 \log G \tag{1.5}$$

the glare index (GI) is for artificial lighting. Introducing a glare index for daylighting derived from the IES glare index for artificial lighting, we obtain the equation:

$$DGI = \frac{2}{3}(\mathrm{GI} + 14), \tag{1.6}$$

for values up to 28, the equation expresses the well-known fact that there is greater tolerance for glare from the sky through windows compared to the same situation with artificial lighting.

Results from studies on the impact of window size and sky luminance in spaces with an average internal reflectance of 0.4 show that the discomfort glare level is nearly constant (±0.5 glare index) for window sizes larger than 25% of the floor area. Therefore, the space exhibits a relatively consistent glare characteristic [67].

Combining vertical illuminance with a modified DGI relationship, Wienold and Christoffersen developed the DGP glare index for daylighting. This metric was developed based on photometric measurements and user feedback in two European cities. The DGP glare index is expressed by the equation:

$$\mathrm{DGP} = 5.87 \cdot 10^{-5} E_v + 9.18 \cdot 10^{-2} \log\left(1 + \frac{L_{s,i}^2\, \omega_{s,i}}{E_v^{1.87} P_i^2}\right) + 0.16, \tag{1.7}$$

where E_v – vertical Eye illuminance (lux),
$L_{s,i}$ – luminance of source (cd/m^2),
ω – solid angle source (–),
P – guth position index.

The DGI and DGP glare indices are well-known metrics used for assessing daylighting conditions in spaces illuminated by large daylight sources such as glazed facades. However, DGP provides better results, especially in bright daylight scenes, and correlates more strongly with subjective discomfort glare. It tends to yield more reliable outcomes compared to other visual comfort indices [68].

Wienold later proposed a simplified version called the simplified glare index for daylighting (DGPs), which considers only vertical illuminance and maintains a strong correlation. It's important to note that the DGPs model is accurate but does not account for direct sunlight. It is expressed by the equation:

Table 1.1 Comparison of glare indices

Glare criterion	DGI	VCP	UGR	DGP	UGP
Imperceptible	Below 18	80–100	Below 13	Below 35	Comfort below 0,5
Perceptible	18–24	60–80	13–22	0.35–0.40	Discomfort above 0,5
Disturbing	24–31	40–60	22–28	0.40–0.45	Discomfort above 0,5
Intolerable	Above 31	Below 40	Above 28	Above 0.45	Discomfort above 0,5

$$\text{DGPs} = 6.22 \cdot 10^{-5} E_v + 0.184, \tag{1.8}$$

Among other glare indices are: Visual Comfort Probability (VCP), Unified Glare Rating (UGR), Daylight Glare Probability (DGP), and Unified Glare Probability (UGP). Glare indices with their respective acceptable ranges are listed in Table 1.1 [65, 69].

In 1995, the Unified Glare Rating (UGR) system was introduced to assess glare in indoor workspaces. This unified system for evaluating glare from lighting systems is also incorporated into standards governing the principles of proper design for indoor lighting systems, such as STN EN 12464-1 Lighting of work places—Part 1: Indoor work places. Designers are required to use this methodology in their designs. It is established that the higher the UGR value, the greater the degree of glare to which individuals are exposed from the indoor lighting system [67, 70–72].

Important measures to reduce visual discomfort include:

- reducing sky luminance,
- reducing the contrast between the sky and its surroundings,
- reducing the visibility of the sky by obstacles.

The brightness of the sky can be reduced by fixed or adjustable devices. Fixed devices such as blinds, louvers, overhangs, etc., restrict daylight entry under all circumstances, some being applied due to low glass transmissivity. Moving shades, curtains, awnings, and similar solutions are highly advantageous when their positioning is adjusted according to sky conditions. Light-colored diffuse walls and ceilings, which produce high brightness, reduce contrast and glare [71, 72].

Another method to reduce discomfort from glare is to increase the brightness of the walls surrounding the window, which decreases contrast. This effect can be achieved by artificial lighting on the adjacent window walls. Similar effects can be achieved using light-colored diffuse window frames and sills [67, 73].

Daylight systems can be categorized in various ways, depending on the context and criteria such as system types, functions, or implementation methods. The latest trends and technologies in daylighting often involve advanced systems designed to maximize energy efficiency and user comfort. Here is an overview of the latest and most commonly used types of daylight systems [65]:

1. Passive Daylighting Systems
Passive daylighting systems do not require additional energy inputs and are designed to optimize natural light through building design.

- Windows: The basic element of daylighting that allows direct access to light into the interior,
- Light Tubes or Sun Tunnels: These transport light from the exterior to the interior using reflective surfaces,
- Atrium: Large open spaces, often with glazed roofs, bringing light into the center of buildings,
- Double-Skin Facades: These allow controlled flow of light and heat between the outer and inner layers of the facade,
- Light Wells: Openings or cavities in roofs and walls that allow light to penetrate deep into the building.

2. Active Daylighting Systems
Active systems utilize technology to monitor, control, and optimize the transmission of daylight.

- Dynamic Shading: Systems that automatically adjust shading (e.g., blinds or curtains) based on sunlight intensity and building orientation,
- Light Shelves: Horizontal elements placed above windows that reflect daylight deeper into the interior while reducing glare,
- Smart Glass: Glass that can change its optical properties (e.g., transparency or reflectivity) based on an electric signal, temperature, or solar radiation,
- Solar Tracking Systems: Devices that track the sun's position and optimize the orientation of mirrors or panels to maximize light transmission.

3. Hybrid Daylighting Systems
These systems combine elements of passive and active technologies to achieve optimal results.

- Integrated Lighting Systems: Systems that combine natural daylight with artificial lighting, using sensors and control systems to maintain constant lighting conditions,
- Shaded Skylights: A combination of skylights with dynamic shading to control the amount and quality of light entering the building.

4. Advanced Materials and Technologies
Innovations in materials and technologies also contribute to the efficient use of daylight.

- Nanotechnology and Photonic Materials: Materials that control and manipulate light on a very fine level, allowing better distribution and control of daylight,
- Photonic Crystals: Materials that selectively filter and reflect certain wavelengths of light, enabling better control of lighting conditions.

The development of daylighting systems is moving towards greater integration with smart technologies, automation, and the use of renewable energy sources. The goal is to maximize energy efficiency, improve user comfort, and minimize environmental impact. These systems are designed to enhance indoor environmental quality, reduce energy consumption, and increase building sustainability. The choice of a specific system or combination of systems depends on the architectural and functional requirements of the project.

1.7 Regulation of Daylighting in Buildings

One definition of visual comfort is that it is a subjective state of visual well-being induced by the visual environment. It depends on the physiology of the human eye, physical quantities describing the amount of light and its distribution in space, and the spectral emission of the light source.

Visual comfort is commonly assessed as the relationship between human needs and the lighting environment (amount of light, uniformity of light, quality of light in rendering colors, prediction of glare risk for users). Research indicates that through regulation, building energy consumption can be reduced by up to 30% [4, 74–76].

Suggestions for regulating glare can be summarized as follows [4, 23, 77, 78]:

- Exposure of glazing on the southern facade for regulating direct sunlight can be achieved by using indirect, diffuse radiation through low-emissivity glazing,
- To minimize the amount of glazing on east and west orientations (which have shading issues compared to south-facing windows),
- eliminate unnecessary shading of windows on the northern facade (windows receive very little direct solar heat gain),
- consider daylighting along with shading,
- internal shading devices such as Venetian blinds or vertical slats do not reduce cooling load (solar gain is always present at the workstation)—these devices allow control of glare and can influence visual comfort in the workplace.

1.8 Innovative Shading Systems for Buildings

Currently, with renewed interest in reducing energy consumption and improving lighting conditions, the use of innovative shading systems for buildings is a positive element in building design. The effectiveness of these systems can be evaluated from various perspectives, but primarily according to [34, 54, 75, 79]:

1. bring natural daylight into interior spaces;
2. reducing overheating (thermal load) of interior spaces due to solar radiation.

Among innovative shading systems for buildings, we can include [54, 75, 79, 80]:

- light shelves,
- anidolic ceilings,
- prismatic glazing,
- laser-cut acrylic panels,
- holographic optical elements (holographic coatings),
- adaptive shading systems (autamated blinds and shades, smart glass, dynamic facades, electrochromic windows, biomimetic Shading Systems, integrated Building Management Systems (BMS).

1.9 Daylighting and Legislation

Normative binding conditions for the assessment and design of interior spaces in terms of daylighting are similar across all European countries. The fundamental characteristic of the lighting climate is the illuminance at the horizontal plane [81, 82].

The International Commission on Illumination (CIE) has standardized two extreme types of sky conditions—overcast sky with graded luminance (known as CIE overcast sky, adopted in 1955) and clear sky (adopted in 1973)—with corresponding relationships to express outdoor horizontal illuminance. In our lighting standards, only the CIE overcast sky with graded luminance has been standardized so far, for both dark and light surfaces [13, 83, 84].

Based on long-term measurements, it has been shown that this type of sky prevails in Slovakia from November to February. To prevent this from affecting the interpretation of calculation results, artificial skies are used. Artificial skies can be found, for example, in Turin, Lausanne, Cardiff, and Limelette.

The advantage of artificial skies is that they always provide stable, comparable outdoor conditions for model measurements, allowing the study of the influences of room geometry, building structures, and daylighting elements on light distribution in interior spaces.

In Slovakia, the artificial sky at the Institute of Construction and Architecture of the Slovak Academy of Sciences (ÚSTARCH SAV) in Bratislava has been in operation since 1968. It features a hemispherical shape with a diameter of 8 m, making it a unique laboratory device for simulating sky luminances. This simulation facility consists of a reinforced concrete dome forming a hemispherical sky, with a special light-diffusing plaster on its inner surface. The dome is illuminated from the floor by scenic luminaires with conventional high-energy consumption bulbs. In the center of the dome, there is a work table for placing room models, sets of objects, or other equipment. The current capabilities of this sky allow simulation of standard CIE sky conditions: uniformly overcast sky, uniform sky, and clear sky [83].

Since 1986, Slovakia has adhered to the core standard STN 73 0580-1, which introduces assumptions about uneven sky luminance distribution with elevation height above the horizon and a theoretically abstracted model of unfavorable winter sky conditions. The reference horizontal illuminance outdoors is specified at 5 000 lx.

The reference area for investigating lighting conditions determined by calculation or measurement is set at 850 mm above the floor in communication spaces at floor level, or directly in the visual task zone in industrial operations where detailed observation is conducted [20].

In order to better assess sky luminance, 15 standards of luminance distribution have been proposed. These standards are designed for international adoption in the CIE/ISO standard 'Spatial distribution of daylight—General sky CIE'. The standard provides procedures for determining the distribution of relative luminance in the sky under typical outdoor conditions. This means it covers not only overcast and clear skies but also includes a new CIE Technical Report (CIE TS) aimed at deriving relationships for calculating luminance distribution and daylight in absolute units. Luminance values are given in cd/m^2 and illuminance in lx [83, 85].

Due to the dynamic nature of daylighting, particularly in Europe, the relative expression of indoor lighting level by daylight factor has become established as a fundamental parameter for assessing indoor lighting quality. However, the new procedure also allows for the assessment of daylight factor and absolute values of luminance and illuminance under typical outdoor skies in real natural conditions and with different atmospheric properties. For the design and evaluation of daylighting in industrial buildings, the Czechoslovak Standard (ČSN) 73 0580-4 'Daylighting of industrial buildings' from 1994 applies [83].

In spaces with lateral and combined lighting where lateral lighting predominates, the average daylight factor value is controlled. In these spaces, this value must not fall below 5% for classes I-III, 3% for class IV, and 2% for classes V-VII of visual tasks. The standard also addresses solutions for lighting openings and appropriate levels of protection for various types of indoor spaces and activities [20].

Currently, methods for evaluating daylighting are not standardized. However, in 2012, preparations began to introduce the first European standard for daylighting and sunlighting of buildings. According to initial proposals, daylighting will be assessed using climate-based criteria. The standard will verify daylighting in buildings and its simulation under artificial sky conditions. This initiative will provide Slovakia with one of the most advanced daylighting simulators globally [13, 84].

At the end of 2018, the first European standard on daylight in buildings, EN 17037, was issued. The goal of this standard is to ensure buildings have quality daylighting. It introduces criteria such as views from buildings, prevention of glare, and sunlight exposure. Several European countries have had norms or regulations recommending minimum daylight levels in specific indoor spaces for several years. The requirements of the standard for ensuring daylight concern indoor spaces of buildings with long-term occupancy by people [13].

According to the standard, a space is considered adequately daylighted if the target daylight factors are achieved at the midpoint of the considered reference plane for at least half of the hours in a year with daylight. Daylight provision in indoor spaces is demonstrated computationally using two methods with validated software: either using daylight factor targets (DT) and minimum daylight factor targets (DTM), or using illuminance targets (ET) and minimum illuminance targets (ETM). Daylight provision is expressed in terms of daylight factors and illuminance in lux. Both methods ensure lighting levels in indoor spaces are expressed in absolute values [13, 84].

The daylight factor method, although fundamentally preserved in the standard (based on the median local exterior diffuse sky illuminance on the horizontal plane), is now considered equivalent to the climate-based daylighting modeling method. In recent years, the climate-based modeling method for daylighting has been favored in many published scientific articles in the field of daylighting. The daylight factor method is considered outdated and surpassed.

Currently, the concept of climate-based simulations for daylighting of buildings has been incorporated into EN 17037 and is even beginning to be applied in building regulations (Denmark). Climate-based modeling of daylighting considers direct sunlight as well [13]. When calculating glare, preference is given to the calculation of the Daylight Glare Probability (DGP) index over the Unified Glare Rating (UGR) index [13].

In the context of climate-based daylight modeling, the most commonly used criterion is spatial daylight autonomy, which is essentially adopted from EN 17037. This criterion evaluates the percentage of the area within a space where a defined target or minimum daylight level is achieved for 50% of daylight hours throughout the year. Neither in LEED v4 (Leadership in Energy and Environmental Design, a sustainability certification system developed in the USA) nor in EN 17037 is there any differentiation made regarding the purpose of the indoor space with respect to daylighting. The same requirement for providing daylight applies to residential rooms, classrooms, offices, hospital rooms, any workplace with prolonged occupancy, etc. [13, 84, 86] (Tables 1.2 and 1.3).

Table 1.2 The values of the daylight factor DF needed to exceed the illumination level 100, 300, 500 alebo 750 lx in Bratislava during 50% daytime hours according to EN 17037 [86]

Illumination level (lx)	DF in Bratislava (%)
100	0.6
300	1.8
500	3.1
750	4.6

Table 1.3 The values of the daylight factor DF required to exceed the illumination levels of 100, 300, 500, or 750 lx in Bratislava during 50% of daytime hours for top lighting apertures filled with light-diffusing material according to EN 17037 [86]

Illumination level (lx)	DF in Bratislava $E_{v,g,med} = 22\,600$ lx (%)
100	0.4
300	1.3
500	2.2
750	3.3

1.10 Methods for Assessing Daylighting

Daylighting of indoor spaces is characterized by the daylight factor. This metric originated from the need to mathematically express the illumination of interiors by skylight. Unlike the sky factor, the actual daylight factor characterizes the overall interior illumination as measurable in real interiors, accounting for light losses and gains as well as specific conditions of objects and interiors (such as shading, reflectivity, dirt, etc.) [4, 31].

Currently, there are several methods and software packages available for assessing year-round daylighting performance, such as Daylight Autonomy (DA) or Model Climate Based Daylight (MCBD). The fundamental criterion in assessing the usability of daylight in the Daylight Autonomy method is the defined percentage of hours per year when interior illumination is expected to be provided solely by daylight.

In contrast to the concept of the daylight factor, the DA approach considers the average illuminance on a surface under all sky conditions, described by diffuse and direct sunlight accessible in the dataset [35, 87].

1.10.1 Computational Methods for Determining the Daylight Factor in Interiors

To comprehend the nuances of daylight characteristics and effectively harness its potential in architectural design, a framework of daylight performance metrics has been proposed. These metrics serve as quality measures for assessing the behavior of daylight within spaces and facilitate comparative analyses across different measures. This approach establishes benchmarks for reference purposes [88].

The performance metrics are categorized into two primary types: Static Daylight Performance Metrics, such as Illuminance (Lux), and Dynamic Daylight Performance Metrics, also referred to as "Climate-Based Daylight Modeling." The latter considers daily and seasonal variations in daylight at a specific building site, taking into account both quantity and quality aspects).

There is ample evidence supporting the efficacy of these metrics, which can accurately predict luminous quantities based on standardized meteorological data tailored to the local climate of the building under evaluation.

The actual daylight factor undergoes changes during the use of an interior and can be referred to as its operational value. During the design stage of a lighting aperture, it's necessary to consider a value that encompasses all influences identified before the construction is completed. From these requirements arises the resultant daylight factor D (hereinafter referred to as DF), expressed as a percentage. This value is determined through calculation under certain simplified assumptions.

It is determined as the sum of components calculated separately, and the resultant daylight factor is determined as the superposition of these components [4, 67]:

$$D = D_S + D_e + D_i = 100 \cdot \left(\frac{E_{is}}{E_{eH}} + \frac{E_{ie}}{E_{eH}} + \frac{E_{ir}}{E_{eH}} \right), \tag{1.9}$$

where

D_S – is the sky component of the daylight factor expresses the ratio of interior illuminance E_{is} (lx), to the current horizontal illuminance E_{eH} (lx) of the external horizontal plane, unobstructed by anything,

D_e – is the external reflective component of the daylight factor expresses the ratio of the interior illuminance caused by light reflected from external reflective surfaces E_{ie} (lx) to the current horizontal illuminance E_{eH} (lx) of the external horizontal plane, unobstructed by anything,

D_i – is the internal reflective component of the daylight factor, expressing the ratio of the interior illuminance caused by multiple reflections of daylight from interior surfaces E_{ir}(lx) to the current horizontal illuminance E_{eH} (lx) of the external horizontal plane, unobstructed by anything.

1.10.2 Simulation Computational Methods

Simulations are a valuable tool for architects, engineers, and developers to analyze and optimize building parameters such as shape, size, orientation, and system type, which influence overall energy consumption of the building [89–91].

The fundamental premise of using computer simulations is to mathematically analyze the given problem. Current simulation programs serve as experimental tools in the realm of virtual reality, allowing for the measurement of various parameters for future buildings.

Computer simulations are employed to assess the built environment at different stages of project preparation, optimize material and construction choices, environmental techniques for thermal and visual comfort, energy consumption, and many other tasks [89, 91–93].

1.10.3 The Application of Computer Simulations

In the world, there are numerous building simulation programs. These simulation programs in a 'real' environment typically consist of interconnected software systems that are most commonly used in practice [91, 94, 95]:

- evaluating visual and thermal comfort in indoor building spaces,
- assessing indoor air quality, concentration of pollutants, smoke spread, and fire behavior,

- predicting building energy consumption for heating, ventilation, and cooling across arbitrary time intervals with detailed information about their internal structure within a room, zone, or entire building,
- determining the annual temperature profiles of air and individual surfaces within various rooms of a building, to analyze the risk of overheating indoor spaces,
- analyzing indoor air humidity and moisture condensation throughout the year, determining shading and insolation of outdoor and indoor surfaces in geometrically complex situations,
- designing 'realistic' regulation and optimizing environmental technology operation in buildings,
- to determine the costs of buildings in terms of their entire life cycle of operation,
- for the rational application of new materials, structural elements, and systems in specific buildings, considering their interrelations (e.g., special glazing systems, sunshades, wind turbines, active and passive solar systems, glazed atria, solar walls, solar collectors, photovoltaic elements, etc.),
- the issue concerning buildings relates to their dynamic behavior (e.g., analyzing the impact of low-emissivity interior and exterior surfaces on their comfort and energy properties).

Three common types of simulations used in architectural design include [4, 96]:

1. building energy simulation—it is the largest and most important application. It includes full and abbreviated hourly intervals, these simulations are used to forecast the energy use of buildings through analysis of heat, ventilation, air conditioning (HVAC), and other factors influencing the energy efficiency of the building;
2. lighting and daylighting simulation—the effects on energy and light efficiency, these simulations focus on analyzing natural lighting and its distribution within building interiors, which can impact visual comfort and the energy costs associated with artificial lighting;
3. solar system simulation—passive and active solar systems, refers to different types of solar energy systems used in buildings. Passive solar systems utilize architectural elements (like building orientation, shading, and materials) to maximize natural heat and light, while active solar systems use technologies (like solar panels and solar water heaters) to convert sunlight into usable energy.

These simulations help architects and engineers better understand and optimize various aspects of building design, leading to more efficient and sustainable solutions.

The application of simulation should be considered based on the nature of the problem and the method of investigation. According to Newton, James, and Bartholomew (1988), there are seven fundamental steps for successful analysis [97–99]:

1. problem definition;
2. model creation;
3. data acquisition;

4. implementation;
5. design;
6. test measurements;
7. analysis of results and final report.

1.11 Partial Conclusion

In a world recently focused on carbon emissions, global warming, and sustainable design, the planned utilization of natural light in buildings has become a crucial strategy for improving energy efficiency by minimizing the burden of lighting, heating, and cooling. The introduction of innovative and advanced daylighting strategies and systems can significantly reduce electricity consumption in buildings and enhance the quality of light in indoor environments.

Daylight, as an element of the environment, is distinguished by its uniqueness, irreproducibility, and perceived naturalness. Its integration into building design not only reduces energy demand but also contributes to a more pleasant and productive indoor environment.

References

1. P. Boyce, C. Hunter, O. Howlett, (2003)
2. V. Bystřický, J. Kaňka, *Osvětlení*, 2nd edn. (ČVUT Praha, 1997)
3. G. Oakley, S. Riffat, L. Shao, Solar Energy **69**(2), 89 (2000). DOI https://doi.org/10.1016/S0038-092X(00)00049-9. URL http://www.sciencedirect.com/science/article/pii/S0038092X00000499
4. E. Dolnikova, D. Katunský, *Hodnotenie svetelnej pohody pri kombinovanom systéme v priemyselných halách* (TU Košice, 2017)
5. I. Acosta, J. Navarro, J.J. Sendra, P. Esquivias, Energy and Buildings **49**, 394 (2012). DOI https://doi.org/10.1016/j.enbuild.2012.02.038. URL http://www.sciencedirect.com/science/article/pii/S0378778812001302
6. CIE-117, Discomfort glare in interior lighting. Tech. rep., Bureau Central de la Commission Internationale de L'Eclairage, Vienna Austria (1995)
7. A. Irakoze, Z.A. Lee, H. Kim, Sustainability **12**(8), 3117 (2020). DOI https://doi.org/10.3390/su12083117. URL http://dx.doi.org/10.3390/su12083117
8. S. Darula. Daylighting. http://www.odbornicasopisy.cz (1999)
9. I. Douglas. New glazing choices for architectural daylighting. http://www.chebucto.ns.ca/Technology/RS2000/papers/0821-01 (2000)
10. L.T. Doulos, A. Tsangrassoulis, E.N. Madias, S. Niavis, A. Kontadakis, P.A. Kontaxis, V.T. Kontargyri, K. Skalkou, F. Topalis, E. Manolis, et al., Energies **13**(15), 4024 (2020). DOI https://doi.org/10.3390/en13154024. URL http://dx.doi.org/10.3390/en13154024
11. J. Habel, *Světelná technika a osvětlování* (FCC Public, 1995)
12. H.I. Hellinga, Daylight and view: The influence of windows on the visual quality of indoor spaces. Ph.D. thesis, Architecture and The Built Environment (2013)
13. J. Hraška, Renewable Energy **73**, 109 (2015). DOI https://doi.org/10.1016/j.renene.2014.06.008. URL http://www.sciencedirect.com/science/article/pii/S0960148114003486

14. J. Hraška, Svetlo **22**(4), 38 (2019)
15. P. Hykš, J. Hraška, *Slnečné žiarenie a budovy*. Edícia stavebníckej literatúry (ALFA, 1990). URL https://books.google.sk/books?id=9aIWAQAACAAJ
16. IESNA. Iesna. iesna–daylighting rp-5-99. Dostupné z: http://www.iesna.org (2001)
17. IESNA. Iesna. iesna–industrial lighting rp-7-91. Dostupné z: http://www.iesna.org (2001)
18. K. Johnsen, *Daylight. A basic human need* (International Energy Agency, IEA, 1999), pp. 9–15
19. K. Johnsen, *Daylight in buildings*. See journal : Solar energy : Vol. 69 (suppl.), nos. 1-6, 2000. special issue (2001)
20. Denné osvetlenie budov, Časť 1: Základné požiadavky. Standard (2000)
21. M. Kischoweit-Lopin, (Proceedings of Solar Energy in Architecture and Urban Planning, 1996)
22. Z. Kong, An investigation of daylighting performance in sidelight spaces. Ph.D. thesis, University of Wisconsin-Milwaukee (2018)
23. N. Ruck, et al, *Daylight in Buildings - A source book on daylighting systems and components* (Lawrence Berkeley National Laboratory, Berkeley (USA), 2000). EPFL LBNL-47493
24. D. Katunský, Analýza svetelných pomerov výrobných hál ľahkého priemyslu pri aplikácii združeného osvetľovacieho systému. Ph.D. thesis, SvF TU Košice (1992)
25. I. Acosta, M. Campano, J. Molina, Journal Engineering Technology **8**(5), 358 (2016)
26. K. Alhagla, A. Mansour, R. Elbassuoni, Alexandria Engineering Journal **58**(1), 283 (2019). DOI https://doi.org/10.1016/j.aej.2019.01.004. URL http://www.sciencedirect.com/science/article/pii/S1110016819300043
27. M.D. Kroelinger, InformeDesign **3**(3), 1 (2014). DOI https://doi.org/10.1016/j.solener.2007.07.003
28. K.M. Kumar, N. Kranthi, International Journal of Recent Technology and Engineering **7**(6S2), 1 (2019). DOI https://doi.org/10.1016/j.solener.2007.07.003
29. E. Dolníková, D. Katunský, Environments **6**(5), 54 (2019)
30. E. Dolnikova, D. Katunsky, S. Darula, Building and Environment **180**, 106976 (2020). DOI https://doi.org/10.1016/j.buildenv.2020.106976
31. I. Acosta, J. Navarro, P. Esquivias, 4th International Congress on Energy Engineering and Management (2011)
32. B. Gosowski, P. Lorkowski, M. Redecki, Thin-Walled Structures **108**, 234 (2016). DOI https://doi.org/10.1016/j.tws.2016.08.015
33. Klahos. Pásové střešní světlíky dodají světlo do každé haly. http://www.klahos.cz/pasove-stresni svetliky.htm (2020)
34. L.S.H. Au, Chapter three–innovative daylighting systems. Tech. rep., Cardiff University: Honours Thesis (2000)
35. T. Mavridou, L. Doulos, Buildings **9**(7), 170 (2019). DOI https://doi.org/10.3390/buildings9070170. URL http://dx.doi.org/10.3390/buildings9070170
36. L.S.H. Au, Chapter two–innovative daylighting systems. Tech. rep., Cardiff University: Honours Thesis (2000)
37. F. Kundracik. Holigilm free software for illumination calculation. Dostupné z: davinci.fmph.uniba.sk/ kundracik1/holigilm/ (2019)
38. M. Kocifaj, S. Darula, R. Kittler, Solar Energy **82**(3), 247 (2008). DOI https://doi.org/10.1016/j.solener.2007.07.003
39. P. Rybár, *Denní osvětlení a oslunění budov*. Technická knihovna (ERA, 2002). URL https://books.google.sk/books?id=xluKAAAACAAJ
40. D. Katunský, E. Dolníková, B. Dolník, Sustainability **10**(3), 655 (2018). DOI https://doi.org/10.3390/su10030655. URL http://dx.doi.org/10.3390/su10030655
41. N. Lechner. Introduction to daylighting: Toplighting. lightforum.com. http://www.lightforum.com/library/design/daylighting/toplighting.html (2002)
42. P. Littlefair, Solar Energy **70**(3), 177 (2001). DOI https://doi.org/10.1016/S0038-092X(00)00099-2. Urban Environment
43. S. Selkowitz, E.S. Lee, in *Daylighting 98: Conference Proceedings* (LBNLCanada, 1998), LBNL

44. A. Tsangrassoulis, M. Santamouris, Energy and Buildings **32**(1), 41 (2000). DOI https://doi.org/10.1016/S0378-7788(99)00039-0
45. Y.M.S. Abdelhamid, S.M. Wahba, M. Elhusseiny, Journal of Daylighting **10**(2), 173 (2023). DOI https://doi.org/10.15627/jd.2023.15
46. Daylighting in buildings (1994). Energy research Group, A thermie programme Action.
47. D. McQuillen. The art of daylighting. Dostupné z: http://www.edcmag.com/archives/1-98-13.htm. 28 marec, 2001.
48. A. Nabil, J. Mardaljevic, Lighting Research & Technology **37**(1), 41 (2005). DOI https://doi.org/10.1191/1365782805li128oa
49. F. DeLuca, R. Simson, J. Kurnitski (eds.). *Energy and daylighting performance design of skylights and clerestories in a large hall retail building*, vol. 2 (CLIMA 2016–proceedings of the 12th REHVA World Congress Aalborg University Denmark, 2016)
50. J. O'Conner, E.S. Lee, F.M. Rubinstein, S.E. Selkowitz, Tips for daylighting with windows: The integrated approach. Tech. rep. (1997)
51. D. Phillips, *Daylighting: natural light in architecture* (Elsevier, 2014)
52. E. Piotrowska, A. Borchert, E3S Web Conf. **14**, 01029 (2017). DOI https://doi.org/10.1051/e3sconf/20171401029. URL 10.1051/e3sconf/20171401029
53. m M hasan, H.Y. Chong, A.K. Dutta, EcoLibrium **17**(5), 44 (2018)
54. S. Rayaz, S. Rubab, Materials Science Forum **760**, 79 (2013). DOI https://doi.org/10.4028/www.scientific.net/MSF.760.79
55. E.D. Resources, *Skylighting Guidelines* (2013). URL https://www.buildup.eu/en/practices/publications/skylighting-guidelines
56. C.V. Subramanian, S. Kamalesvari, International Research Journal of Engineering and Technology **3**(12), 1381 (2016)
57. Daylight in buildings. Tech. rep. (1998). Survey Simple Design Tools, IEA SHC Task 21/ECBCS Annex 29
58. N. Dogne, A. Choudhary, International Journal for Scientific Research and Development **3**(9), 130 (2015)
59. C. Pierson, J. Wienold, M. Bodart, Buildings **8**(8), 94 (2018). DOI https://doi.org/10.3390/buildings8080094. URL http://dx.doi.org/10.3390/buildings8080094
60. J.Y. Suk, Building and Environment **148**, 107 (2019). DOI https://doi.org/10.1016/j.buildenv.2018.10.058
61. A. Faraji, F. Rezaei, P. Rahnamayiezekavat, M. Rashidi, H. Soleimani, Sustainability **15**(15) (2023). DOI https://doi.org/10.3390/su151511885
62. M. Khanmohamadi, M. Pourahmadi, F. Mozaffar, Journal of Sustainable Architecture and Urban Design **7**(1), 113
63. O. Labaj, M. Frýbort, *Průmyslové stavby IV. Tvorba pracovního prostředí v průmyslu*. Učební texty vysokých škol (VUT, 1985). URL https://books.google.sk/books?id=i5CeHAAACAAJ
64. J. Rucińska, A. Trząski, Energies **13**(10), 2555 (2020). DOI https://doi.org/10.3390/en13102555. URL http://dx.doi.org/10.3390/en13102555
65. J.W. Clotilde Pierson, M. Bodart, LEUKOS **14**(3), 111 (2018). DOI https://doi.org/10.1080/15502724.2018.1428617
66. M. Bodart, C. Cauwerts, Building and Environment **113**, 210 (2017). DOI https://doi.org/10.1016/j.buildenv.2016.08.033. Advances in daylighting and visual comfort research
67. N. Baker, A. Fanchiotti, K. Steemers, *Daylighting in architecture* (James and James London, 1993)
68. K. Chaloeytoy, M. Ichinose, S.C. Chien, Buildings **10**(10), 180 (2020). DOI https://doi.org/10.3390/buildings10100180
69. J.L. Hensen, R. Lamberts, *Building performance simulation for design and operation* (Routledge Publisher London, 2019)
70. F. Vajkay, Optimisation of light conditions in buildings. Ph.D. thesis, Vysoké učení technické v Brně, Fakulta stavební, Stavební fakulta ÚPS (2012)
71. Leukos, The Journal of the Illuminating Engineering Society **7** (2011)

72. G.J. Ward, F.M. Rubinstein, R.D. Clear, SIGGRAPH Comput. Graph. **22**(4), 85 (1988). DOI https://doi.org/10.1145/378456.378490
73. R. Perez, J. Michalshy, R. Seals, Journal of the Illuminating Engineering Society **21**(2), 84 (1992). DOI https://doi.org/10.1080/00994480.1992.10748005
74. L. Beltran, E. Lee, S. Selkowitz, Journal of The Illuminating Engineering Society **26**, 91 (1997)
75. J. Hraška, *Tieniaca technika budov: typológia, navrhovanie a hodnotenie*, 1st edn. (SPEKTRUM STU Bratislava, Vazovova 5 Bratislava, 2020)
76. *Install a system of light shelves and shading* (1999). Energy efficiency manual
77. G. Courret, J.L. Scartezzini, D. Francioli, J.J. Meyer, Energy and Buildings **28**(1), 79 (1998). DOI https://doi.org/10.1016/S0378-7788(97)00066-2
78. I.R. Edmonds, Solar Energy Materials and Solar Cells **29**(1), 1 (1993). DOI https://doi.org/10.1016/0927-0248(93)90088-K. URL http://www.sciencedirect.com/science/article/pii/092702489390088K
79. M. Kischkoweit-Lopin, Solar Energy **73**(2), 77 (2002). DOI https://doi.org/10.1016/S0038-092X(02)00036-1. Daylighting
80. E. Lee, D. DiBartolomeo, S. Selkowitz, Journal of the Illuminating Engineering Society **28**(1), 3 (1999). DOI https://doi.org/10.1080/00994480.1999.10748247
81. S. Darula. Normovanie požiadaviek na denné osvetlenie a insoláciu interiérov. http://www.odbornicasopisy.cz (2015)
82. S. Darula, Osvetl̆ovacia technika **4**, 4 (1999)
83. S. Darula, R. Kittler, H. Kambezidis, A. Bartzolas. Guidelines for more realistic daylight exterior conditions in energy conscious designs. Computer adaption and examples, Bratislava, Atény (2000)
84. S. Darula, VTS news **1**(6), 16 (2018)
85. R. Kittler, S. Darula, A set of standard skies. Final report, Institute of Construction and Architecture, Bratislava (1998)
86. Daylight of building. Standard (2018)
87. F.D. Luca, M. Kiil, J. Kurnitski, R. Murula. Evaluating daylight factor standard through climate based daylight simulations and overheating regulations in estonia (2019). DOI https://doi.org/10.13140/RG.2.2.23057.07525
88. A.A.I. Nessim, Journal of Engineering Sciences **45**, 380 (2017). DOI https://doi.org/10.21608/jesaun.2017.116276
89. J. Clarke, J. Hraška, M. Janák, Stavebná ročenka pp. 97–101 (1999)
90. D. DiLaura, D. Igoe, P. Samaras, A. Smith, Journal of the Illuminating Engineering Society **17**(1), 36 (1988). DOI https://doi.org/10.1080/00994480.1988.10748706
91. J. Hraška, M. Janák, Architektúra, Stavebníctvo, bývanie **7**, 60 (2000)
92. S. Herkel, in *Building Simulation '97, Vol.II., 5.international IBPSA Conference* (IBSA Prague, 1997), pp. 307–312
93. M. Janák, EUROSTAV **6**, 29 (2000)
94. A. Khodulev, E. Kopylov, *Physically accurate lighting simulation in computer graphics software* (Keldysh Institute of Applied Mathematics, 1996). Russian Academy of Science, Moscow State University
95. J. Love, M. Navvab, Journal of the Illuminating Engineering Society **20**(1), 140 (1991). DOI https://doi.org/10.1080/00994480.1991.10748933
96. E. Ng, K. Lam, W. Wu, in *Building Simulation '99 Proceedings*, vol. 2 (Kyoto Japan, 1999), vol. 2, pp. 555–562
97. J. Mardaljevic, Daylight simulation: Validation, sky models and daylight coefficients. Ph.D. thesis, De Montfort University, Leicester, UK (1999)
98. E. Vartainen, Daylight modelling with the simulation tool delight. Tech. rep., Helsinki (2000). Report TKK-F-A 799
99. G.J. Ward, F.M. Rubinstein, Journal of the Illuminating Engineering Society **17**(1), 80 (1988). DOI https://doi.org/10.1080/00994480.1988.10748710

Chapter 2
Simulation Methods for Daylight

Daylight is a better choice for replacing artificial lighting during the day, which can result in lower energy consumption for lighting. In building design, computer simulation tools are very useful analytical tools that can assist in evaluating architectural designs related to daylighting and energy assessment. There are several simulation tools available for performing energy and daylighting analysis in buildings under various sky conditions.

2.1 Principle of Simulation Methods

Simulation methods for daylighting are based on so-called first principles, from classical optical, thermodynamic, or lighting models of radiation propagation. The fundamental methods used in computer simulations of lighting technology are: the *Monte Carlo* method, which applies the ray tracing technique (reverse ray tracing), and the radiosity method [1–5].

Monte Carlo method works with very small surfaces in a stochastic manner (results of repeated calculations of the same problem may slightly differ). The ray tracing method is a rendering technique based on calculating the distribution of a large number of rays emitted into a scene, either from light sources (forward ray tracing) or from the viewpoint (backward ray tracing).

Backward ray tracing is a faster method than forward ray tracing because it only computes rays directed toward the viewpoint. However, it is less suitable for scenarios where it is difficult to locate light sources with narrow dimensions, such as light pipes.

Ray tracing algorithms support reflective, transmissive, and refractive properties of surfaces, allowing the use of complex materials in simulations. The backward ray tracing method is used by the simulation program RADIANCE [3, 5–7].

© The Author(s), under exclusive license to Springer Nature Switzerland AG 2025

E. Dolníková, *Simulation Tools for Predicting Daylighting Conditions in Buildings*, Digital Innovations in Architecture, Engineering and Construction,
https://doi.org/10.1007/978-3-031-78330-2_2

The ray tracing technique is explained by the integral equation for each point of the model:

$$L_r(\theta_r, \phi_r) = L_e(\theta_r, \phi_r) + \iint L_i(\theta_i, \phi_i)\, \rho_{bd}(\theta_i, \phi_i; \theta_r, \phi_r)\, |\cos\theta_i|\, \sin\theta_i\, d\theta_i\, d\phi_i\,, \tag{2.1}$$

where θ is polar angle measured from the surface normal,

ϕ	– azimuthal angle measured around the surface normal,
$L_e(\theta_r, \phi_r)$	–emitted radiance ($W \cdot sr^{-1} \cdot m^{-2}$),
$L_r(\theta_r, \phi_r)$	–reflected radiance ($W \cdot sr^{-1} \cdot m^{-2}$),
$L_i(\theta_r, \phi_r)$	–incident radiance ($W \cdot sr^{-1} \cdot m^{-2}$),
$\rho_{bd}(\theta_i, \phi_i; \theta_r, \phi_r)$	–bidirectional reflectance distribution function and transmission (sr^{-1}).

The radiosity method works with larger surfaces deterministically (results of repeated calculations of the same problem are always the same). The radiosity method includes an algorithm capable of rendering realistic scenes with shadows and diffuse light. It is an application of the finite element method to solve the rendering equation for scenes with purely diffuse surfaces, a method originally developed for studying heat transfer. This method requires the surfaces of the scene to be divided into a mesh of smaller patches.

The method has conditions that limit its use for daylight simulations (diffuse surfaces, complex sky description) and should only be used to evaluate relatively simple spaces. This method is used by simulation programs Relux, DIALux, and LightCalc [3, 5].

There is also a method—Photon Mapping, that is a rendering technique utilizing bidirectional ray tracing. The first pass combines forward ray tracing and photon mapping to distribute light in the scene and balance the light flux on surfaces. The computation of the final image then utilizes a second pass consisting of backward ray tracing from the viewpoint. This optimized technique allows for accurate simulation of more complex lighting scenarios, such as light tubes, and is faster than traditional backward ray tracing.

The simulation program VELUX Daylight Visualizer employs the Photon Mapping method [3, 5]. An illustration of the basic principles of three main algorithms can be seen in Fig. 2.1 [8].

For effective application of global illumination simulation in buildings, their combined use is advantageous. Simulation software is applicable for daylight analysis and building energy assessment. By employing simulation software, 80% of daylight is scrutinized. Simulation results are influenced by the sky condition; a significant portion of the software utilizes CIE sky models. Under various climatic conditions, it is necessary to define the brightness distribution in the sky, as brightness integration into windows is computed [9].

Daylight simulations are highly recommended in the initial design phase when proposed elements are still being defined. This requires many assumptions about

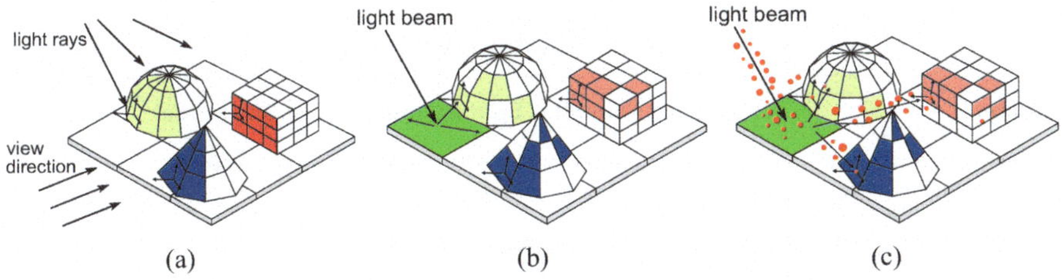

Fig. 2.1 Diagram illustrating the basic principles of three common methods used in daylight simulation: **a**—ray tracing method, **b**—radiative method, **c**—photon mapping [8]

input values, which will significantly influence the outputs and require careful professional interpretation. For instance, the selection of surface reflectance can result in ±60 % differences in output for some daylight metrics. Also, the choice of simulation method affects the results.

Research has shown that there was a difference of ±15% in results using different illuminance-based techniques. Daylight simulations, therefore, are not a simple supplement to daylight theory but rather a discipline in itself. It requires a depth of understanding of inputs and their impact on daylight.

Knowledge of the theory and methods used by lighting simulation software is also important (for example, why the radiative method may not be suitable for glare assessment or why the same simulation may provide slightly different results when conducted using different software [9].

2.2 Software for Lighting Environment Simulations

Simulation programs are a useful tool for designing and developing energy-efficient lighting systems in buildings and for predicting daylighting in building interiors. There are many free and commercial programs available on the market, such as DaySim, RADIANCE, Ecotect, DiaLux, VeLux, DOE-2, EnergyPlus, etc.

These programs can be combined; for example, a building model imported within the Ecotect program can be transferred to EnergyPlus for energy consumption reduction simulations [10].

Simulated results generated by a given simulation program are compared with measured values. The shading effect should be evaluated during the initial building design. Figure 2.2 shows a general scheme of a daylight simulation program [10].

Daylighting is not just a matter of energy savings, as it also concerns human health. Daylight creates a positive environment in the workplace and provides users with a sense of comfort, which automatically enhances their productivity and ability to think. Designers propose various simulation models that would be helpful in further developing daylight design in buildings [5, 6].

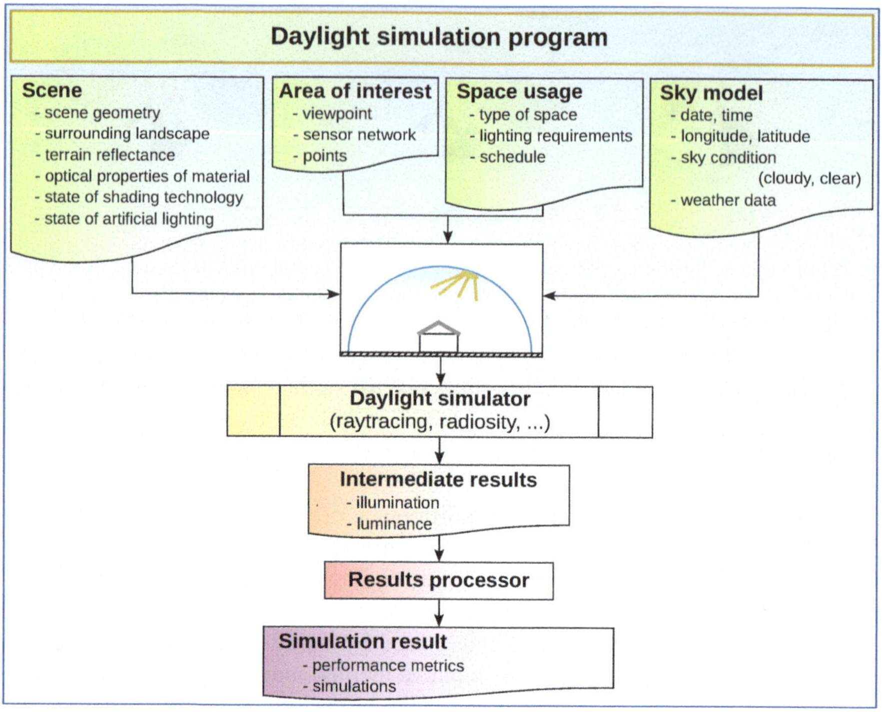

Fig. 2.2 Diagram of the daylight simulation program [10]

The best software for daylight simulation in the world is widely used and recognized for its accuracy and robustness. Some of the most well-known and widely used software includes: Radiance is a powerful simulation system for predicting and visualizing lighting in a space. It is widely recognized for its accuracy in simulating both daylight and artificial lighting. Advantages: High accuracy, ability to simulate complex lighting conditions, extensive community support. DAYSIM is a simulation tool for annual daylight analysis based on Radiance. It is used to predict daylight in buildings and analyze lighting based on user preferences. Advantages: Focuses on long-term daylight analysis, integration with CAD tools, support for dynamic shading systems. Honeybee (part of Ladybug Tools) is a tool for lighting simulation and analysis that uses Radiance and integrates with Grasshopper and Rhino. It is part of the larger Ladybug Tools ecosystem. Advantages: Easy integration into CAD workflow, large user community, support for multiple simulation tools. Velux Daylight Visualizer is a tool specifically aimed at daylight simulation and analysis of lighting conditions in buildings. Advantages: Intuitive user interface, quick and easy simulations, specialization in daylight. DIALux is a comprehensive lighting design software that supports simulation of both daylight and artificial lighting. It is widely used in the lighting industry. Advantages: Intuitive interface, support for

Table 2.1 Overview of daylight simulation programs

Description	RADIANCE	Daysim	Velux D. Visualizer
Type of calculation	b. raytracing[a]	b. raytracing[a]	Photon mapping[c]
Outputs	DF, E	DF, E	DF, E
Average value of DF	Yes	Yes	Yes
General interface	Very difficult	Slightly difficult	Very easy
Device geometry	Very difficult	Slightly difficult	Very easy
Importing model	Yes	Yes	Yes
File types	.obj, .skp., .dxf, .dwg	.eco, .skp, .3ds, .rad	.obj, .skp., .dxf, .dwg
Graph outputs	Average	Very difficult	Very easy
Description	DIAlux	Ecotect	LightCalc
Type of calculation	radiosity[d]	BRE Split Flux	f. raytracing[b]
Outputs	DF, E	DF, E	DF, E
Average value of DF	No	áno	áno
General interface	Easy	Very easy	Average
Device geometry	Average	Easy	Very easy
Importing model	Yes	Yes	Yes
File types	.sat, .skp., .m3d, .3ds	.obj, .dwg, .dxf	.skp
Graph outputs	Easy	Very easy	Very difficult

[a] Backward ray tracing
[b] Forward ray tracing
[c] Photon mapping
[d] Radiation method

various types of lighting, wide practical use. IES VE (Integrated Environmental Solutions Virtual Environment) is comprehensive software for building simulation and analysis, including daylight, energy efficiency, and environmental impacts. Advantages: Integrated environment for various types of analysis, wide functionality, high accuracy.

Choosing the best software depends on the specific needs and requirements of the project, as well as the user's level of expertise. Radiance and its extensions (like Honeybee and DAYSIM) are often preferred for their accuracy and ability to simulate complex lighting conditions, while tools like Velux Daylight Visualizer and DIALux are known for their user-friendliness and speed of simulations.

An overview of some significant simulation programs for daylighting calculation, along with their characteristics, is provided in Table 2.1 [3, 6].

2.3 The Simulation Program RADIANCE

Radiance is a powerful simulation system for predicting and visualizing lighting in a space, enabling the creation of realistic models of lighting conditions. Honeybee is a plugin for visualization and simulation tools like Grasshopper (a visualization plugin for Rhino), integrating Radiance, EnergyPlus, and other simulation tools into one workspace. In architecture and construction, Radiance and Honeybee are used for light and lighting simulations and analyses. By combining these two tools, Radiance and Honeybee, architects and designers can accurately model and analyze lighting conditions in their designs. This allows them to optimize lighting, improve building energy efficiency, and ensure that the indoor environment is pleasant and healthy for its users.

The development of the simulation program RADIANCE began sometime in the mid 1980s at the Lawrence Berkeley Laboratory in the USA. For over three decades, Greg Ward has been the main developer and maintainer of this program. RADIANCE is a physically-based rendering program and has been validated by numerous empirical studies. Mardaljevic (1999) was also the first to comprehensively document and validate RADIANCE for daylighting calculations using daylight coefficients.

The daylight coefficient method, according to its original purpose, enables efficient lighting calculations for various sky conditions through matrix computations [5, 6, 11].

2.3.1 *Characteristics of the Program*

The model used in RADIANCE contains thousands of surfaces, created separately in CAD programs. It models planar polygons, spheres, and cones, as well as more complex shapes such as cubes and prisms [5, 6].

Each surface is defined by orientation. Objects forming the "scene" are expressed in a Cartesian coordinate system (X, Y, Z). The X-axis points eastward, the Y-axis northward, and the Z-axis toward the zenith. Individual vertices are specified by unit lengths. Each simple ray is characterized by the amount of radiance expressed in $\mathrm{W\,m^{-2}\,sr^{-1}}$. Irradiance is divided into three "channels": red (r), green (g), and blue (b). The total irradiance R is calculated as the sum of irradiance components R_r, R_g, and R_b [5, 11]:

$$R = 0,263\, R_r + 0,655\, R_g + 0,082\, R_b \tag{2.2}$$

where

$$0,263 + 0,655 + 0,082 = 1\,.$$

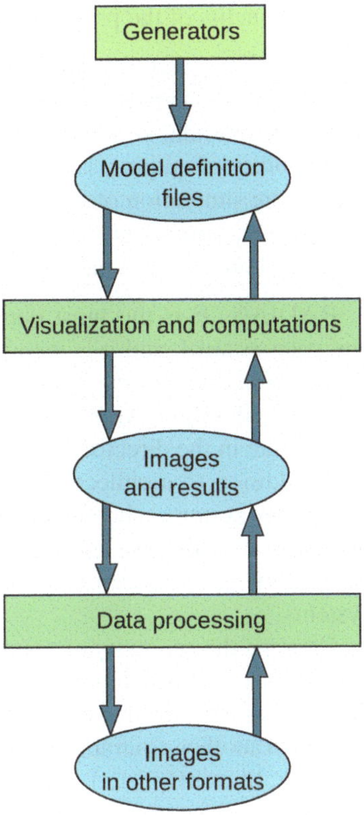

Fig. 2.3 Block diagram of the simulation program RADIANCE

Irradiance transformation R (radiometric unit) to luminance L (photometric unit) is given by the equation:

$$L = 179\,R = 47,1\,R_r + 117,2\,R_g + 14,7\,R_b\,. \tag{2.3}$$

The structure diagram of the RADIANCE program is in the Fig. 2.3.

RADIANCE is a free tool for lighting environments that engineering firms extensively utilize for innovatively managing solar radiation, illumination, and daylighting to enhance building energy efficiency. RADIANCE offers complete flexibility concerning scene geometry and materials and has been validated through detailed measurements.

In recent years, advancements in radiation have led to the ability to perform annual simulations in less than two minutes and significantly improve the accuracy in modeling anisotropic systems and materials that scatter light, common in many shading and daylighting devices. The RADIANCE simulation program continues to improve to support fast and accurate annual simulations for complex fenestration systems, including coplanar and non-coplanar shading systems [9].

The benefits of the program include:

- free access, reliability, and continuous development,
- collaboration between users and the program's creator,
- flexible modular structure for various issues,
- variability of outputs,
- accurate simulation of light behavior in complex conditions,
- correct computational results,
- nearly faithful reproduction of a photograph.

RADIANCE is the most commonly used tool in lighting design, daylighting, and solar control, and it is incorporated into numerous free and commercial architectural and engineering applications, including IES Virtual Environment, Design Builder, and Sefaira Architecture. A dynamic list of applications using RADIANCE is available in the directory of building energy software tools by IBPSA.

In addition to architects and engineers, RADIANCE is used by manufacturers of window systems, shading devices, and daylighting for the development and optimization of both new and existing lighting, daylighting, and solar control products, allowing for the simulation of the performance of a wide range of fenestration systems.

Creating the geometry of a simple space in the RADIANCE program can take less than 15 min, and simulation time can range from one to two minutes to several hours depending on the desired calculation accuracy [5]. The model's geometry can also be created in another program and imported into the RADIANCE program (e.g., in CAD software, SketchUp, etc.). Simulation time depends on the computer's performance as well as factors [6, 12, 13]:

- the number of light sources and reflections,
- image size,
- the accuracy of rendering parameters.

The program then saves the calculated result values to a file and reuses them for subsequent creation of the same model for different perspectives, but in a shorter time. To accurately evaluate the final result of lighting environment analyses, a certain level of experience is required [3, 4].

RADIANCE has become the foundation of architectural and lighting design companies due to its widespread adoption and well-validated results. RADIANCE is a collection of software programs, not an application, and the RADIANCE ecosystem includes three types of programs. They are:

- the basic RADIANCE programs,
- various derivative works,
- different graphical user interfaces.

RADIANCE employs backward ray tracing, where primary rays emanate from a point (a virtual camera or a light sensor) into the environment sample. Whenever a ray intersects a surface, it recursively generates one or more new rays depending on the surface material and accumulates their results into a single value, which is then returned as the result of the parent ray.

For direct and specular reflections, a small number of rays are typically required to be generated, whereas a much larger number of rays are generated for sampling indirect illumination due to environmental lighting at the intersection. Consequently, calculations of the environment tend to dominate the overall ray tracing computation time. In RADIANCE, each ray returns red, green, and blue (RGB) values in radiation units. The array of values returned from primary rays produces sensor image values.

The second class of tools consists of derivative works based on RADIANCE, which modify their source code to add new functionality. For example, the simulation program DaySim is a modification of the RADIANCE program that calculates daylight coefficients instead of red, green, and blue channels for each ray. The Accelerad program belongs to this second class of programs.

For the creation of the Accelerad program, the source code of RADIANCE was modified to incorporate the OptiX™ GPU ray tracing library developed by NVIDIA (an American company manufacturing processors for graphics cards and chipsets for motherboards). The Accelerad application is designed to accept the same command line interface to enable these user interfaces to define and substitute the original RADIANCE and DaySim programs [5, 14–17].

2.4 The Simulation Program Honeybee

Honeybee is a software tool and plugin used for simulations and analyses in building physics and environmental design. It is part of a larger ecosystem known as Ladybug Tools, which includes several tools for environmental analysis, energy simulation, and building optimization. Honeybee integrates and facilitates the use of several powerful simulation engines, including EnergyPlus, Radiance, and OpenStudio. Key features of Honeybee [18–20]:

- Lighting Simulation: Using Radiance, Honeybee can simulate daylight and artificial lighting in buildings, enabling the analysis and optimization of lighting conditions,
- Energy Simulation: With EnergyPlus and OpenStudio, Honeybee can simulate building energy consumption, aiding in the design of energy-efficient buildings,
- Environmental Analyses: It allows for various types of analyses, such as thermal comfort, indoor environmental quality, and other factors affecting occupant comfort and health,
- Integration with Grasshopper: Honeybee functions as a plugin for Grasshopper, a visualization tool for Rhino that enables parametric and visual design, simplifying the creation and modification of models.

Thus, Honeybee is a key tool for architects, engineers, and designers focused on creating ustainable and energy-efficient buildings, providing sophisticated tools for analyzing and optimizing various aspects of construction projects.

2.5 The Simulation Program Evalglare

Evalglare is a software tool used for assessing glare in interior environments. It is part of the Radiance package, which is widely used for lighting simulations. Evalglare allows for the analysis of how different lighting conditions affect visual comfort in a space. Key features of Evalaglare [18–20]:

- Glare Detection and Analysis: Evalglare identifies glare sources in images (HDR—High Dynamic Range) and calculates various glare indices, such as Daylight Glare Probability (DGP), Glare Index (GI), and Unified Glare Rating (UGR),
- Simulations and Measurements: Evalglare can simulate and measure glare under different conditions, aiding in the design of lighting systems and the optimization of architectural elements to minimize glare,
- Result Visualization: The tool allows for the visualization of analysis results, making it easier to interpret and present lighting design outcomes.

Thus, Evalglare is a valuable tool for architects, engineers, and designers focused on ensuring visual comfort and effective lighting in building interiors. It helps identify and address glare-related issues, contributing to the creation of comfortable and productive environments for users.

2.6 The Simulation Program Accelerad

The Accelerad program is a free set of tools for fast, accurate analysis, and visualization of natural and artificial lighting. Originally developed by Nathaniel Jones at the MIT Sustainable Design Lab, it continues to be independently developed. The program is modeled after the standard software package RADIANCE developed by G. Ward at the Lawrence Berkeley National Laboratory. To facilitate smooth adoption between RADIANCE users and Accelerad software developers, Accelerad maintains compatibility with RADIANCE file formats, materials, and command line arguments [21, 22].

Accelerad is designed to mimic the behavior of the RADIANCE program and has been tested compared to Radiance 5.2. While there are numerous other sources for Radiance executable files, including DIVA, DaySim, and IES VE, these distributions may not be updated as frequently and have not been tested using the Accelerad program [21, 22].

Thanks to parallelism and RTX architecture, Accelerad is up to a hundred times faster than RADIANCE, leveraging OptiX™, a ray tracing technology based on the graphics processing unit (GPU). Accelerad is ideal for simulations that would be impractical to run on processors due to their size, such as rpict rendering with a large number of ambient reflections or rtrace calculations on many thousands of sensor points.

The new compatibility with rcontrib allows for fast simulation of annual daylight metrics based on climate through three-phase and five-phase simulations [21].

AcceleradRT, as the next generation of daylight simulation, represents an interactive interface for real-time daylighting, glare, and visual comfort analysis. The AcceleradRT application utilizes progressive path tracing to ensure real-time daylight simulation results with verified accuracy.

This tool creates visualizations along with data on lighting levels and visual comfort, automatically responding to changes in underlying geometric and material models. It includes AcceleradVR, an immersive visualization interface compatible with most virtual reality add-on kits [21, 22].

2.7 The Simulation Program LightStanza

LightStanza is an advanced software tool for daylight simulation and analysis, used in architecture, construction, and the lighting industry. It is a cloud-based platform that allows users to perform complex lighting simulations directly in a browser without the need for software installation. Key Features and Benefits of LightStanza [19, 20]:

- Simulation of Daylight and Artificial Lighting—LightStanza enables accurate daylight simulations that aid in designing energy-efficient and comfortable indoor spaces, users can also simulate artificial lighting and analyze its interaction with daylight,
- Intuitive User Interface—The platform features a user-friendly and intuitive interface, making it accessible even to less experienced users, simulations can be performed directly in the browser without the need for complex software installation and configuration,
- Cloud Architecture—Thanks to its cloud architecture, LightStanza leverages the computational power of remote servers, allowing for quick and efficient execution of complex simulations, users can access their projects from any device with an internet connection,
- Comprehensive Analytical Tools—LightStanza provides detailed lighting analyses, including daylight, workspace lighting, glare analysis, and other important metrics, users can generate reports and visualizations to aid in communicating results with clients and teams,
- Integration with BIM and CAD Software—LightStanza can import and export models to and from popular BIM and CAD software, facilitating integration into existing workflows, supports formats such as IFC, OBJ, and others,
- Sustainability and Certifications—Assists projects in achieving sustainability goals and obtaining certifications such as LEED, WELL, and others, analyzes how building designs meet the requirements of these certifications, particularly in the area of natural lighting.

LightStanza is used by architects, engineers, lighting designers, and other construction professionals to optimize building designs and improve indoor environmental quality. It enables them to quickly and efficiently perform complex lighting simulations, identify potential issues, and optimize designs to meet the highest standards of energy efficiency and user comfort. LightStanza is a modern and powerful tool for daylight and artificial lighting simulation, known for its accessibility, accuracy, and ability to integrate with other software tools. Its use leads to better-designed buildings that are energy-efficient and provide a high-quality indoor environment.

2.8 The Simulation Program Velux Daylight Vizualizer

The VELUX Daylight Visualizer is a simple, validated tool for simulating and visualizing daylight for the design and analysis of daylight in buildings. It was created by Luxion and the VeLux group. It is intended to support the use of daylight in buildings and helps professionals predict and document the level and appearance of daylight in a space before building project implementation.

It allows for the rapid generation of 3D models into which roof and facade windows can be freely inserted. The program also enables users to import 3D models generated by CAD programs into the Windows system to facilitate a smooth workflow and provide flexibility in model geometry.

The program can be used to calculate illuminance, luminance, and the daylight factor for all 15 types of sky defined in the CIE Standard General Sky. Simulation results are provided in the form of images and numerical values. This program has a good workflow and is easy to use, attracting both beginners and experienced designers. The program is freely accessible [9].

2.9 Partial Conclusion

Currently, computing technology allows for the collection of a larger amount of data in real-time, their processing, analysis, and the acquisition of new knowledge. These facts can be utilized to predict the future behavior of lighting comfort in buildings.

The primary advantage of RADIANCE over simpler lighting calculation and rendering tools is that there are no limitations regarding the geometry or materials that can be simulated. Architects and engineers use RADIANCE to predict lighting, visual quality, and the appearance of innovative design spaces, while researchers assess new lighting technologies and daylighting. RADIANCE is continually updated.

References

1. Leukos, The Journal of the Illuminating Engineering Society **7** (2011)
2. M. Janák, EUROSTAV **6**, 29 (2000)
3. J. Mardaljevic, Daylight simulation: Validation, sky models and daylight coefficients. Ph.D. thesis, De Montfort University, Leicester, UK (1999)
4. J. Mardaljevic, K. Lomas, in *Building Simulation 95, Fourth International Conference Proceeding* (Madison, Winsconsin, 1995), pp. 14–16
5. G.J. Ward. Radiance tutorial. Available from: http://radsite.lbl.gov/radiance/refer/tutorial.html (2001)
6. E. Dolnikova, D. Katunský, *Hodnotenie svetelnej pohody pri kombinovanom systéme v priemyselných halách* (TU Košice, 2017)
7. M. Janák, (zborník Budova a energia Herľany, 1999), pp. 213–218
8. C.E. Ochoa, M.B. Aries, J.L. Hensen, Journal of Building Performance Simulation **5**(4), 209 (2012). DOI https://doi.org/10.1080/19401493.2011.558211
9. A. Ahmad, A. Kumar, O. Prakash, A. Aman, Materials Science for Energy Technologies **3**, 679 (2020). DOI https://doi.org/10.1016/j.mset.2020.07.002
10. J.L. Hensen, R. Lamberts, *Building performance simulation for design and operation* (Routledge Publisher London, 2019)
11. R. Compagnon. Radiance: A simulation tool for daylighting systems. University of Cambridge Department of Architecture Cambridge (1997)
12. G.J. Ward, F.M. Rubinstein, R.D. Clear, SIGGRAPH Comput. Graph. **22**(4), 85 (1988). DOI https://doi.org/10.1145/378456.378490
13. M. Janák, in *Building Simulation '97, Vol.II., 5.international IBPSA Conference* (IBSA Prague, 1997), pp. 313–319
14. D. DiLaura, D. Igoe, P. Samaras, A. Smith, Journal of the Illuminating Engineering Society **17**(1), 36 (1988). DOI https://doi.org/10.1080/00994480.1988.10748706
15. A. Khodulev, E. Kopylov, *Physically accurate lighting simulation in computer graphics software* (Keldysh Institute of Applied Mathematics, 1996). Russian Academy of Science, Moscow State University
16. E. Ng, K. Lam, W. Wu, in *Building Simulation '99 Proceedings*, vol. 2 (Kyoto Japan, 1999), vol. 2, pp. 555–562
17. J. Mardaljevic (ed.). *Validation of lighting simulation program: a study using measured sky brightness distributions* (Lux Europa 97 Proceedings Amsterdam, 1997)
18. M. Khanmohamadi, M. Pourahmadi, F. Mozaffar, Journal of Sustainable Architecture and Urban Design **7**(1), 113
19. J.W. Clotilde Pierson, M. Bodart, LEUKOS **14**(3), 111 (2018). DOI https://doi.org/10.1080/15502724.2018.1428617
20. M. Bodart, C. Cauwerts, Building and Environment **113**, 210 (2017). DOI https://doi.org/10.1016/j.buildenv.2016.08.033. Advances in daylighting and visual comfort research
21. N. Jones. Accelerad, accelerad–daylighting simulation on the gpu. https://nljones.github.io/Accelerad, (2020/11/19) (2020)
22. N.L. Jones, in *Building Simulation 2019, 16th International Conference of the International Building Performance Simulation Association* (IBPSA Italy Roma, 2019), pp. 982–989

Chapter 3
Daylight Measurement in Buildings

In selected buildings, test measurements of daylight, surface reflectance, and sky luminance were conducted to compare the results with computer simulations RADIANCE a Velux Daylight Visualizer with the measured values in a real, furnished interior.

The sky luminance in the case studies was measured on selected days under uniformly overcast skies, with the luminance gradient of the sky measured in all directions and at various elevation angles 15°, 45°, and 90°.

3.1 Measurement of Sky Luminance

The measurement of sky luminance distribution at the location of comparative illuminance was conducted at the beginning and end of the measurement. The measured values were compared with the standardized values of the luminance distribution of a uniformly overcast sky for the given terrain.

The measurement of sky luminance was carried out according to ČSN 36 0011 Daylight Measurement. A luminance meter of type HAGNER UNIVERSAL PHOTOMETER, Model S2, with an accuracy of ±5%, and a calibration sample of HAGNER REFLECTION REFERENCE type with a reflectance factor value of 0.966, a LS-110 photometer with accuracy ±2%. In Fig. 3.1, the measured brightness values of the sky are provided in the case study of a production hall in Košice. From the provided graphs, it can be seen that the measured brightness values were within the prescribed elevation heights of the sky in almost an acceptable range [1].

Glare can be caused by instances where direct sunlight enters a room. A very common source of such glare is direct or reflected sunlight within the observer's field of view, as well as high luminance contrasts. Therefore, it is necessary to eliminate high

© The Author(s), under exclusive license to Springer Nature Switzerland AG 2025

E. Dolníková, *Simulation Tools for Predicting Daylighting Conditions in Buildings*, Digital Innovations in Architecture, Engineering and Construction,
https://doi.org/10.1007/978-3-031-78330-2_3

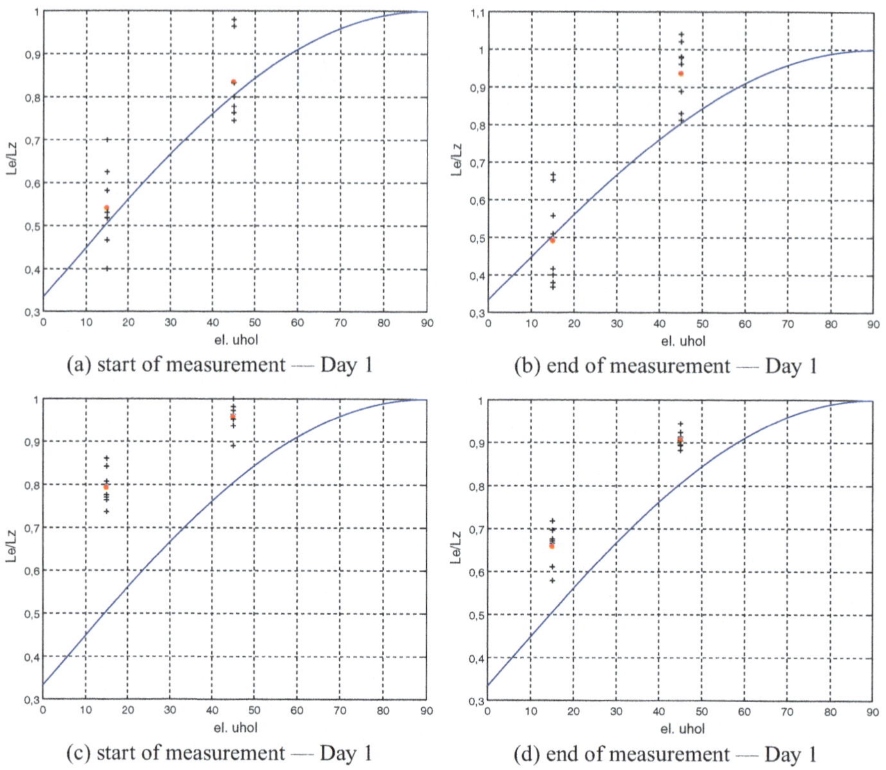

Fig. 3.1 The course of sky brightness gradation distribution in two days of measurement DF

luminance values from openings and surfaces within indoor spaces. It is important to assess the luminance distribution in these areas to ensure optimal comfort during work.

The resulting luminance can be evaluated, for example, based on criteria such as the maximum allowable luminance in the observer's field of view between the observed object and the background, distant dark and light objects, glare index value, as well as the ratio of the luminance of the observed object to the sky.

According to EN 17037, glare from lateral lighting openings is evaluated using the probability of glare method based on eye illuminance in the vertical plane by daylight and the simultaneous brightness of the dazzling source, taking into account its size and position [2].

The risk of glare increases with low eye illuminance and the presence of objects with high luminance in the field of view. The luminance of windows should be controlled within the typical field of view direction at an angle of 60° deviation on each side [1].

The luminance of the lighting opening itself should not exceed the value set by the standard, which is 4,000 cd m^{-2} [3].

In the work, visual comfort was also assessed using glare indices such as GuthVCP, DGP, UGR. The glare index VCP expresses how many percent of people would feel comfortable in a given situation. The value of 70% is typically considered acceptable.

3.2 Measurement of Daylight Illumination

Measurements of internal daylighting were conducted on various days with the artificial lighting system turned off at least twice at each point. The arithmetic average was determined from the obtained values. Control points for measurements were mostly at a height of 850 mm from the floor, as visual performance was needed at that height. Measurements were conducted at control points simultaneously with two luxmeters. The exterior horizontal illumination of the unshaded plane was measured at the level of the roof structure at a height of 1.5 m. Daylight illumination measurement was carried out according to the ČSN 36 0014 Daylight Illumination Measurement standard. Luxmeters of the HAGNER EC 1, ALMEMO FLA 623 VL, and ALMEMO 2690-10A types were used as measuring instruments.

When assessing the availability of daylight, the procedure followed the STN 73 0580-1, EN 12464 standard in functionally defined parts of the space where conditions are set for the values of the daylight factor, uniformity of daylight, light distribution, glare prevention, light reflectance, and surface coloration.

The given object was classified into one of the seven lighting technology classes, which are characterized by visual activity and the proportional observation distance to the critical detail. When assessing daylighting, the following requirements are demanded [4]:

1. the minimum value of the daylight factor,
2. the average value of the daylight factor only for top lighting and combined lighting, provided that the proportion of top lighting to the average value of the daylight factor is at least one half,
3. the uniformity of daylighting in the space with combined lighting is fulfilled when the minimum and average values of the daylight factor are met.

3.3 Measurement of Surface Reflectance

The reflectance factor was determined by calculation according to the STN 36 0014 standard. The brightness of the examined surface L_1 and the brightness of the calibration sample L_2 placed on the surface were measured using a photometer, whose reflectance factor ρ_2 was laboratory verified. The measurement involved determining the brightness of wall surfaces, ceiling, floor, and internal fixtures. From the measured brightness values of individual surfaces and the known reflectance factor of the calibration sample, the resulting reflectance factor of the measured surfaces was calculated using the equation

$$\rho_1 = \rho_2 \frac{L_1}{L_2} . \tag{3.1}$$

The values of the calculated factors suddenly represent average values from multiple measurements. In addition to meeting the requirement of measuring surface reflectivity according to the specified standard, these calculated values were used as input data for a simulation program when assigning surface material properties in the given object [1].

3.4 Measurement of Glazing Properties

The pollution factor was implemented by directing the photometer perpendicularly to the considered window pane, where a portion of the glass is cleaned so that two vertical stripes, one clean and one polluted, are adjacent to each other, against a background with the same and constant brightness. Then, by quickly and parallelly moving the photometer, the brightness of these stripes is read, and pollution is determined according to the equation

$$t_z = \frac{L_z}{L_č} , \tag{3.2}$$

where L_z is the brightness of the polluted glazing,

$L_č$—the brightness of the clean glazing.

Similarly, the glazing transmission factor is determined, with measurements taken with the lighting opening open. The calculated values were again used as input data for the simulation program [1].

3.5 Partial Conclusion

Testing measurements of daylighting in buildings serve as input data for simulation programs. Based on the measurements, which constitute boundary conditions, a model is created. The prerequisite for conducting the simulation is thus the creation of a model. Simulation represents, in simplified terms, an experiment with the model. Simulation tools are part of building design as well as daylighting design for buildings.

References

1. Měření denního osvětlení. Standard (2006, Zrušená 2014)
2. Daylight of building. Standard (2018)
3. J. Hraška, *Tieniaca technika budov: typológia, navrhovanie a hodnotenie*, 1st edn. (SPEKTRUM STU Bratislava, Vazovova 5 Bratislava, 2020)
4. Denné osvetlenie budov, Časť 1: Základné požiadavky. Standard (2000)

Chapter 4
Simulation of Visual Comfort in Buildings

Simulation of daylighting in buildings analyzes daylight and the effects of daylighting on energy and lighting efficiency. In order to create a model for calculation, experimental measurements of daylighting were conducted in selected buildings, which were used as boundary conditions. Subsequently, it was possible to create various other models of objects.

4.1 Creation and Evaluation of the Model

Simulation means to mimic, to simulate. We capture the simulated system in the form of a computer model and then observe what happens in the model. The model of an object is just a simplified representation of the real system, but modeling has many advantages. A model can analyze the behavior of the system in a much shorter time. Processes that take several years in reality can be shortened to fractions of seconds through simulation.

With models, it's easy to manipulate and subsequently conduct experimental calculations by changing their parameters. This allows avoiding interventions in real systems that could jeopardize them. Comparing the costs of realizing a model, which are not negligible, the costs of experimenting with a real system are always lower.

Verification refers to confirming whether a model represents a conceptual model with a specified degree of accuracy. Validation indicates whether the model exhibits a satisfactory degree of agreement with reality in line with its intended use. Evaluation of object models is done through statistical analysis of measured and calculated values.

© The Author(s), under exclusive license to Springer Nature Switzerland AG 2025

E. Dolníková, *Simulation Tools for Predicting Daylighting Conditions in Buildings*, Digital Innovations in Architecture, Engineering and Construction, https://doi.org/10.1007/978-3-031-78330-2_4

4.1.1 Case Studies of Hall Structures

In the individual case studies, models of the respective objects were created based on measurements. The availability of daylighting was evaluated by quantitative assessment parameters: daylight factor (%), illuminance (lx), and uniformity of lighting (–). In a production hall in Košice (*Hala1*), which is oriented longitudinally in the east–west direction with combined lighting, daylighting was analyzed during overcast conditions (Fig. 4.1a). Internal illuminance was measured on a horizontal reference plane at a height of 0.85 m. Additionally, the individual factors of reflectance and transmittance were also measured.

Visual comfort under clear sky conditions was also evaluated. The side lighting system is formed by windows on each cardinal direction. The top lighting system consists of a ridge skylight with wire glass (Fig. 4.1b). A hall model with the original skylight (Fig. 4.2a, b) and an alternative model with an arched skylight (Fig. 4.2c) were created using the simulation program RADIANCE. Daylight factor values for both the ridge and arched skylights were calculated from the measured data, see Table 4.1.

The course of the daylight factor for the ridge skylight—measured values are shown in Fig. 4.3a and simulated values in Fig. 4.3b. The distribution of daylight factor for simulated values of the arched skylight is depicted in Fig. 4.2d. The results of visual comfort assessment are presented in Figs. 4.4a, 4.5a for wired glazing and Fig. 4.4b for diffused glazing.

The results demonstrate that glare occurs even in the case of diffused glazing (values of glare sources are visible in the images), and illuminances are not within the acceptable permissible range. In the hall, simulations of additional skylight models (saw-tooth, lantern, monitor) were conducted, followed by evaluation of daylighting and visual comfort.

The simulation also evaluated the impact of beams and equipment, the contribution of side and overhead lighting to visual comfort, and diffused glazing in the case of the ridge skylight. Only some parameters and accompanying images are presented in the paper.

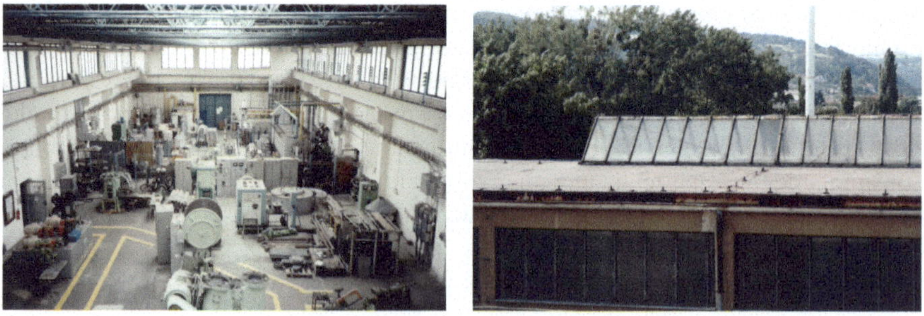

(a) view into the interior (b) view of the ridge skylight

Fig. 4.1 External and internal environment of the hall

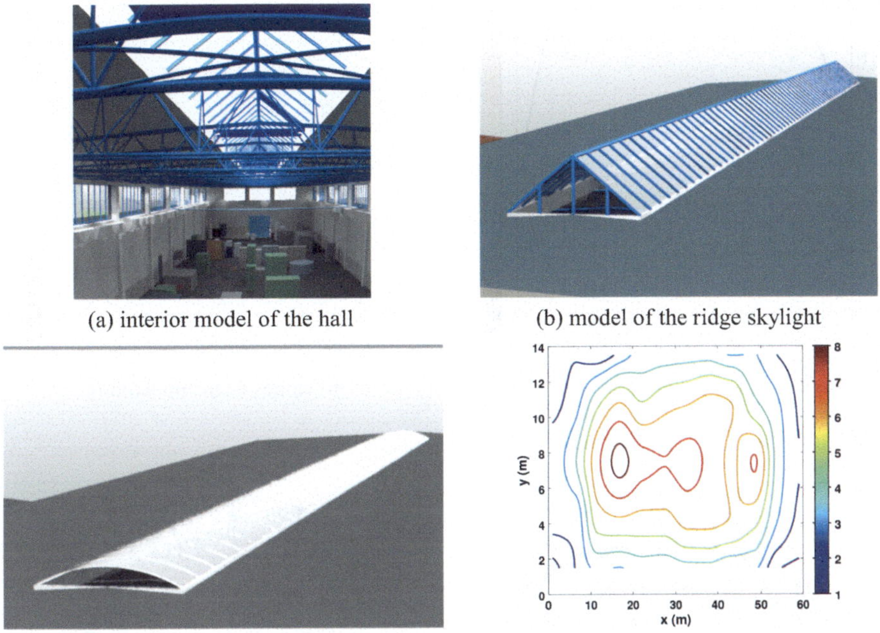

(a) interior model of the hall (b) model of the ridge skylight (c) model of the arched skylight (d) course of daylight illuminance – arch

Fig. 4.2 Models of the hall and skylights

Table 4.1 Assessment parameters for daylighting—Hall 1—before reconstruction

	Daylight factor (%)		
	D_{min}	D_{max}	D_{aver}
Measured	0.44	7.69	4.96
Simulated	2.18	6.96	4.83
Simulated arch	2.88	8.63	6.12

From the results, it is evident that both equipment and beams (more so than equipment) impact visual comfort. Table 4.1 shows that values from measurement and simulation are comparable except for the minimum daylight factor. In the case of the arched skylight, all daylight factor values are elevated.

In evaluating visual comfort using the glare index (Guth VCP), the lowest satisfaction would be with the saddle skylight (10–90%, see Fig. 4.5b), and the highest would be with the sawtooth skylight (100%), depending on the direction of view.

In Hall 1 (*Hala1R*), the skylight was reconstructed into an arched skylight, and the side windows on the northern side were removed. An office with a window on the eastern side was built inside. The position of the equipment—machines remained unchanged. After the reconstruction, measurements were taken again, and the lighting conditions were evaluated using the simulation program RADIANCE.

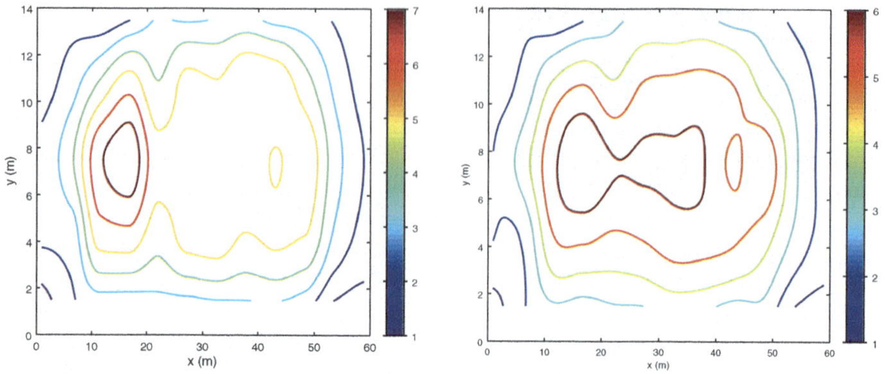

(a) calculated from the measured values

(b) values from simulations

Fig. 4.3 Daylight availability—ridge skylight

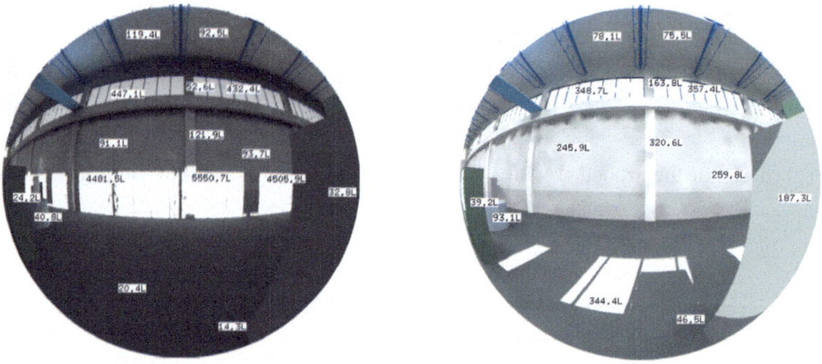

(a) wiber glazing

(b) diffused glazing

Fig. 4.4 Simulation result: clear sky, saddle skylight, north orientation

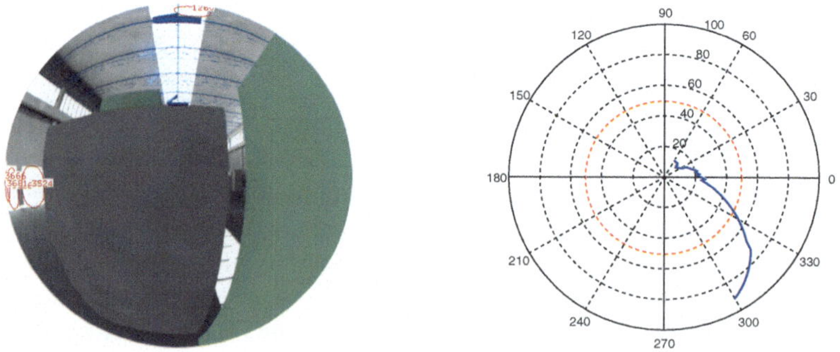

(a) sources of glare, east orientation

(b) glare index

Fig. 4.5 Simulation result: assessment of visual comfort, saddle skylight

Table 4.2 Evaluative parameters of daylighting—Hala1R

	Daylight factor (%)		
	D_{min}	D_{max}	D_{aver}
Measured	1.42	9.57	5.4
Simulated	0.22	7.92	4.15

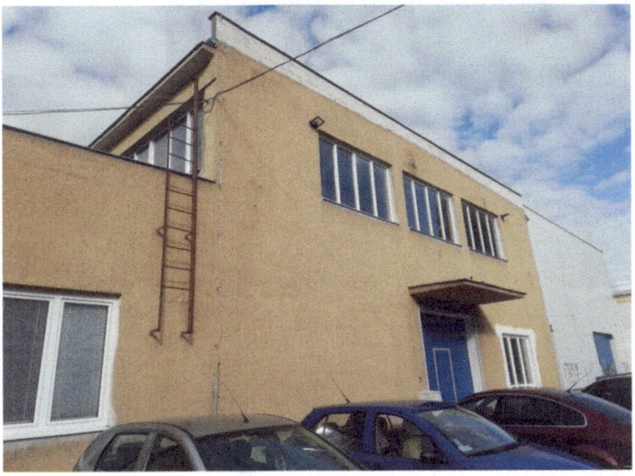

Fig. 4.6 Exterior view of the reconstructed hall

The calculated values of the daylight factor from measured data are shown in Table 4.2. The profiles of daylight factor are depicted in Fig. 4.8. From Table 4.2, it is evident that the largest differences occur at the maximum value. When comparing the simulated daylight factor values before reconstruction with the actually measured values after reconstruction, they differ but are comparable. The discrepancy is also due to the changed boundary conditions (the interior of the hall was modified and side windows were bricked up) which were not considered in the model and calculations before the reconstruction. Figures 4.6 and 4.7 show the interior and exterior of reconstructed Hall 1.

The assessment of daylight conditions was also conducted in a hall of the same type (*Hall 2*), located adjacent to Hall 1, featuring a barrel vault skylight and side windows oriented only to the east-west direction. During measurements, the hall was empty without any internal fixtures. The daylight factor was evaluated from the measured values, as shown in Table 4.3.

Subsequently, based on these measurements, a hall model was created and simulations were performed using the RADIANCE program. The results of the daylight factor profiles are depicted in Fig. 4.10a, b. Figure 4.9 illustrates the brightness values.

From Table 4.3, it is observed that the values are comparable, except for the maximum value. However, even in this hall, the minimum daylight factor requirement is not met. Visual comfort is also compromised, as depicted in Fig. 4.9.

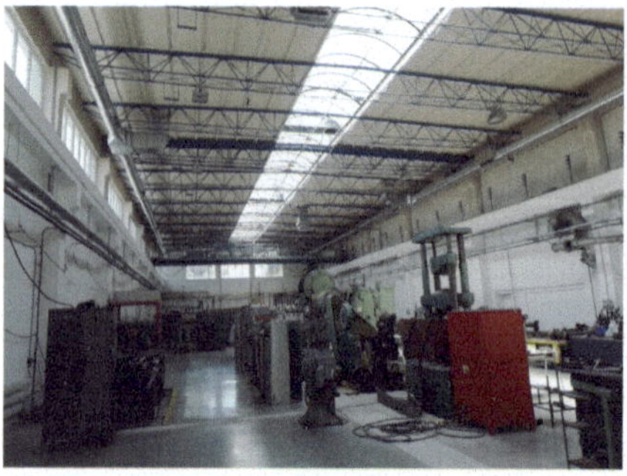

(a) interior view

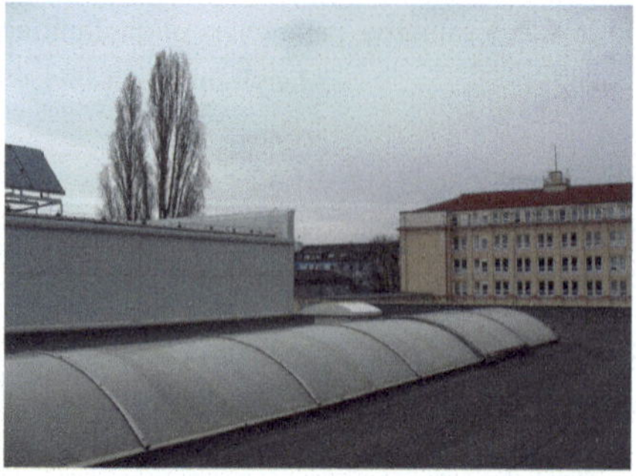

(b) exterior view of the skylight

Fig. 4.7 Reconstructed Hall 1

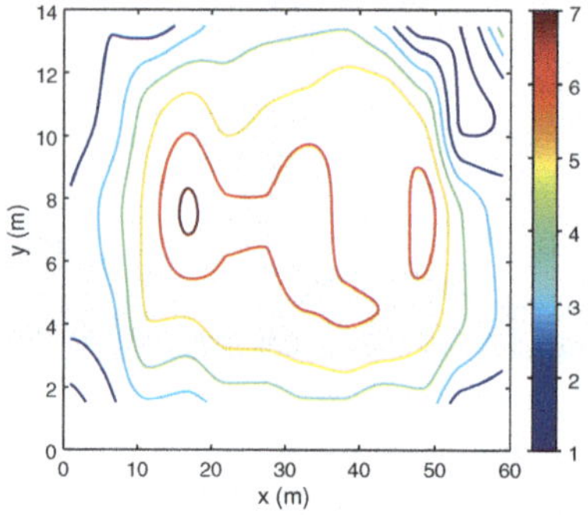

(a) curves of equal illumination

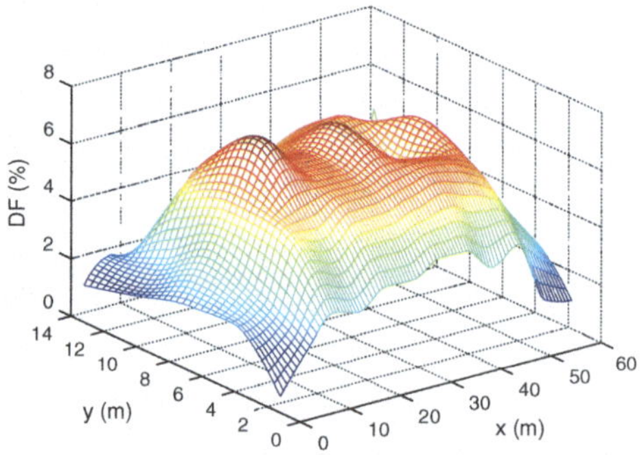

(b) spatial distribution

Fig. 4.8 Profiles of daylight availability—reconstructed Hall 1

Table 4.3 Evaluative parameters of daylighting—Hall 2

	Daylight factor		
	D_{min}	D_{max}	D_{aver}
Measured arch	0.81	10.45	4.63
Simulated	0.99	6.69	3.59

In the mentioned hall, the interior was reconstructed, as shown in Fig. 4.10c, d. Creating a model of the reconstructed hall and subsequently evaluating the lighting conditions will be the subject of further research.

Another case study involves textile manufacturing halls in Košice (*Hall F and Hall G*), as depicted in Fig. 4.11a, b, featuring side daylighting (on adjacent and opposite walls). These halls have dimensions of 54 × 54 m and are equipped with sewing machines. In industrial buildings, window openings typically constitute around 3–6% of the floor area, whereas in residential buildings, windows make up approximately 10% of the floor area.

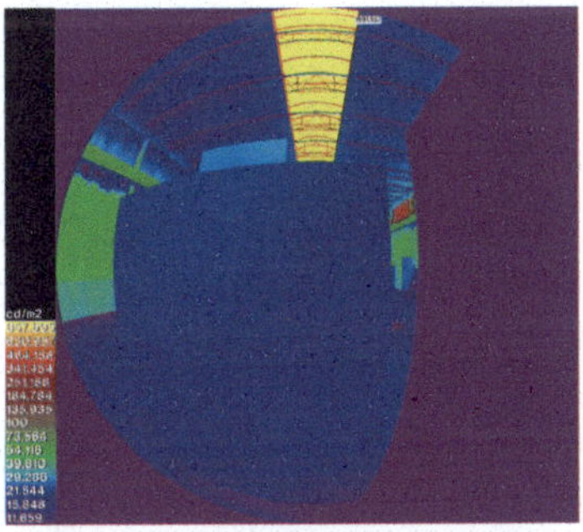

(a) sources of glare

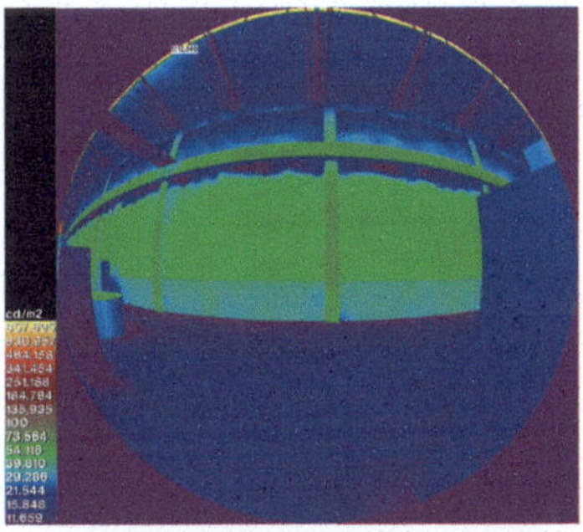

(b) brightness values

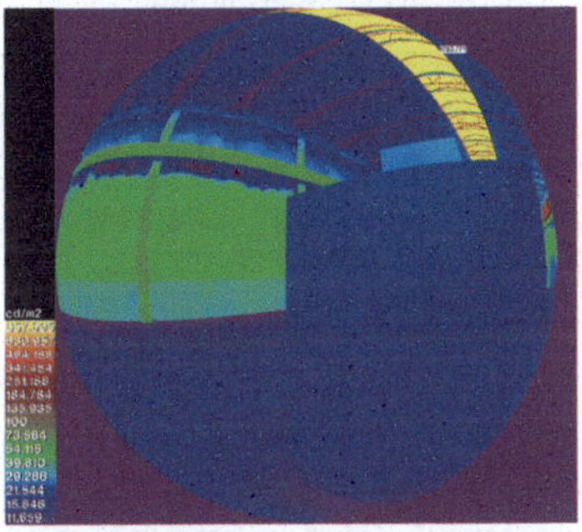

(c) sources of glare

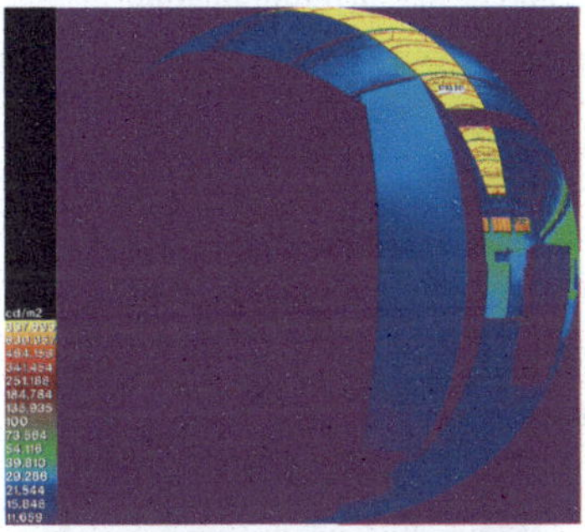

(d) brightness values

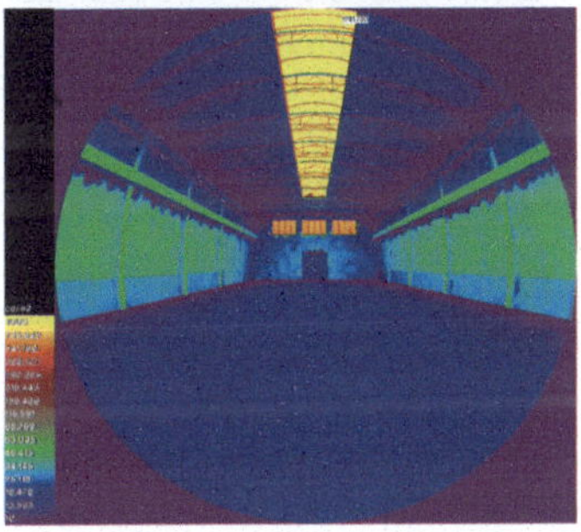

(e) The curves of equal illumination

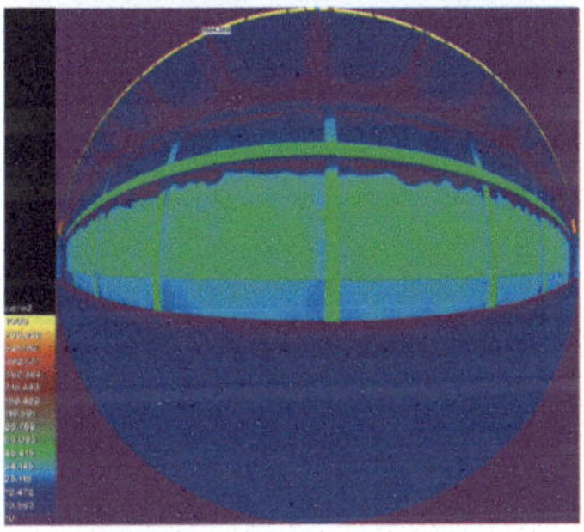

(f) spatial distribution

Fig. 4.9 Simulation result—Hall 2

In the mentioned halls, measurements and calculations of the daylight factor results were conducted, considering the effects of various parameters (such as window size and area), influences of natural and artificial lighting. The following boundary conditions were considered: exterior illuminance of 5 000 lx; hall height of 5.1 m and parapet height of 1.2 m. Window sizes varied (heights of 1 500 mm and widths of 1 800, 2 100, and 2 400 mm) as did the window frame thickness (150, 250, 300, 400 mm). Light characteristics included: light transmittance (92, 74, 80, 100%),

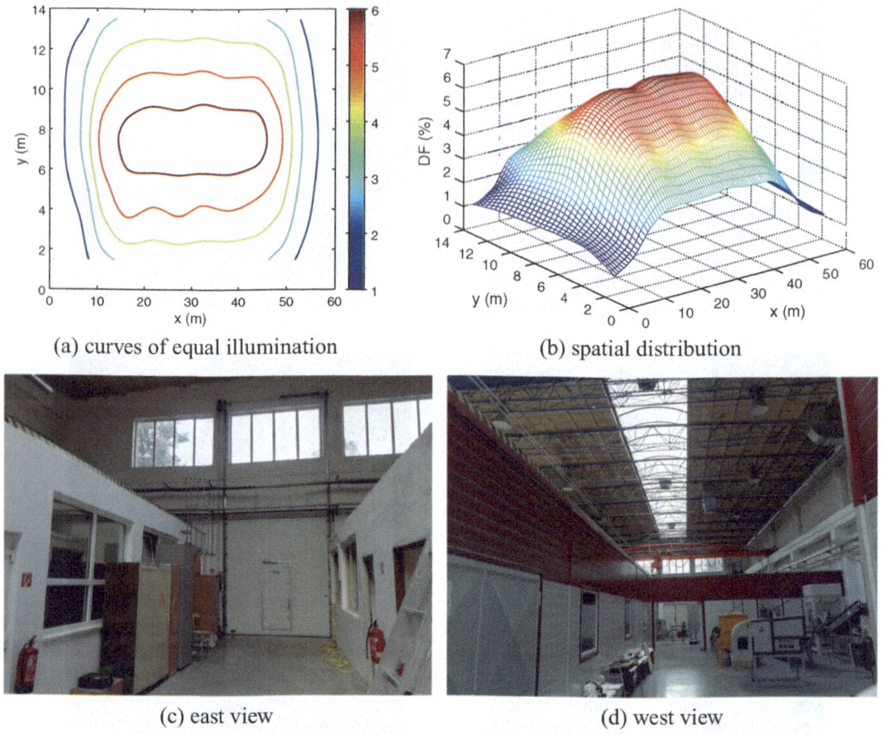

Fig. 4.10 Profiles of daylight availability and interior view—reconstructed Hall 2

surface reflectance: white ceiling—75%, window—10%, walls—70%, dark ground—15%, floor—50%, and facade—45%. Shading obstacles were not considered in the calculations.

From the results, it can be summarized that the daylight factor decreases with the depth of the tract, and the impact of the jamb thickness in a hall with unilateral side lighting becomes significant at about 2 m from the windows. Additionally, as the thickness of the jamb and the sill increases, the sky component decreases. For all window alternatives, the daylight factor varied significantly near the windows and stabilized at a distance of about 15 m.

The minimum values of the daylight factor for all window alternatives met the standards. While the measured and calculated daylight factor values were higher near the windows, they were lower further away from the windows. Artificial lighting was also measured in the absence of daylight. The average illuminance was around 552 lx with a uniformity ratio of 0.63.

The daylighting factor generally increases with larger window areas relative to the total floor area. Therefore, the ratio of window area to floor area has been expressed. Closer to the windows, this ratio depends more on the daylighting factor than at greater distances from the windows. For example, at a distance of 5 m from the

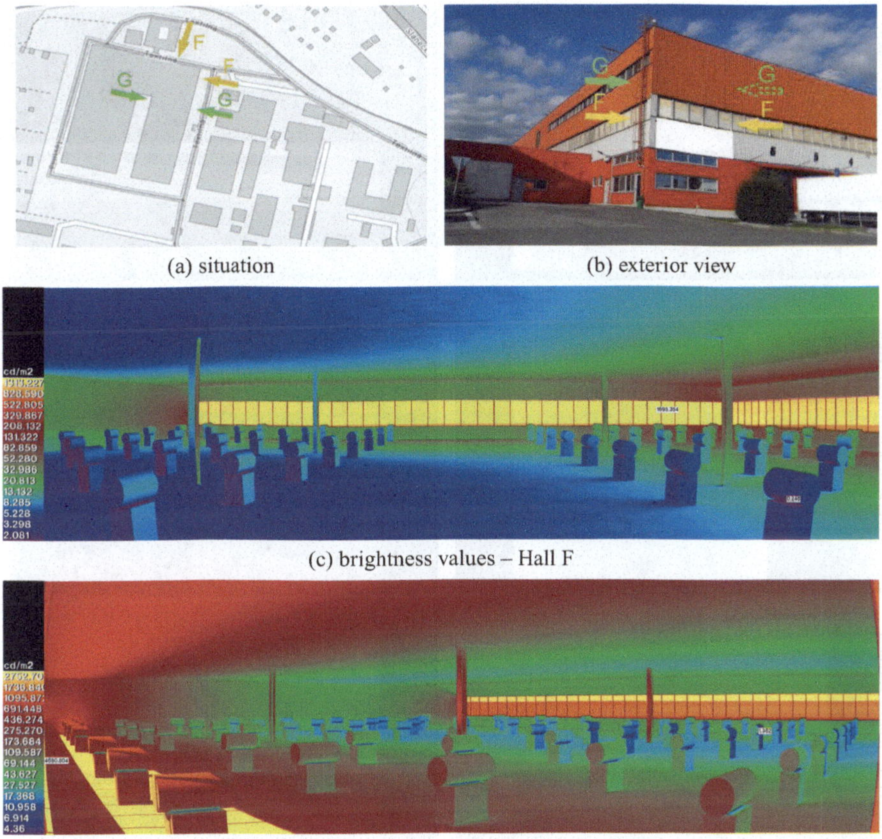

(a) situation (b) exterior view

(c) brightness values – Hall F

(d) brightness values – Hall G

Fig. 4.11 Situation of the textile hall with an exterior view and brightness evaluation

windows, the effect of the window area is almost negligible. At a distance of 9 m, the daylighting factor has almost doubled, and the ratio has increased by 4–6%.

In the study, variations in the reflective component of the daylighting factor were also discussed in relation to distance from the windows. When the glazed area of the window was increased by raising its height by 300 mm, the daylighting factor increased by approximately 30%, and its reflective component increased on average by 5–10%. Upon examining more distant points, it was found that the measured values of the daylighting factor closely matched its sky component. This sky component was on average 30% higher than the measured values of the daylighting factor. In large industrial halls, using a side lighting system, values of the daylighting factor can be determined with sufficient accuracy by multiplying the sky component of the daylighting factor by a coefficient of 1.3 at points deeper than 9 m from the windows.

It was found that natural daylight in the investigated hall does not meet the required conditions throughout the working space. In points and areas where the light levels

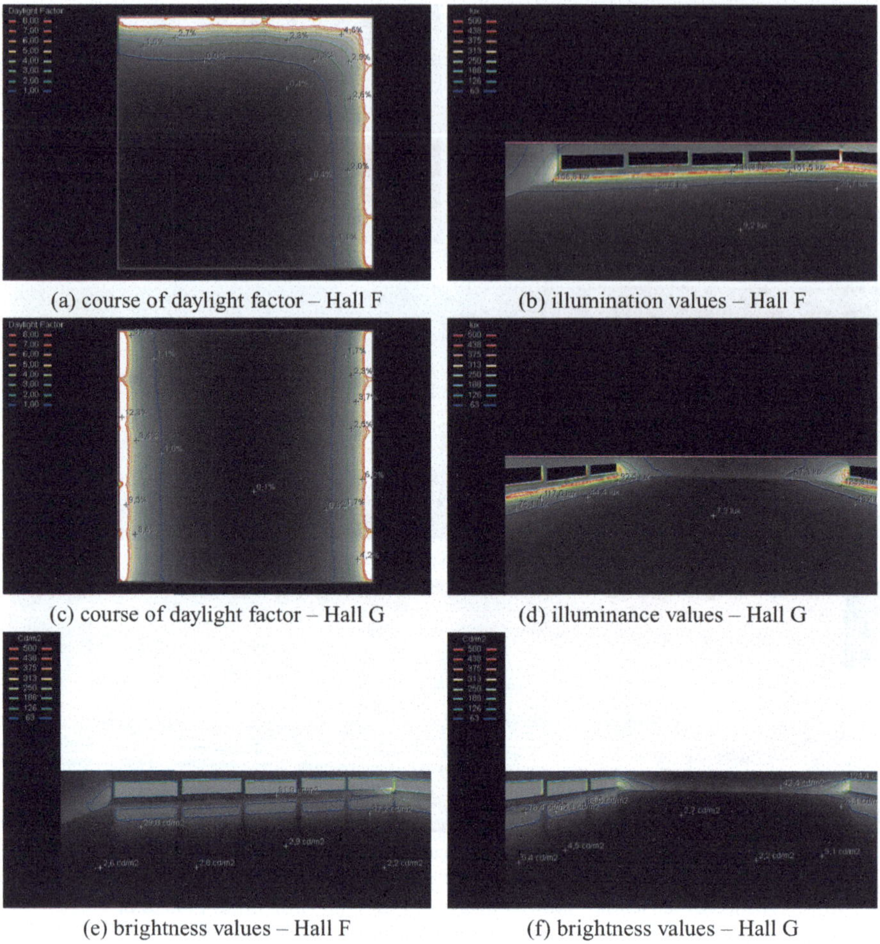

(a) course of daylight factor – Hall F
(b) illumination values – Hall F
(c) course of daylight factor – Hall G
(d) illuminance values – Hall G
(e) brightness values – Hall F
(f) brightness values – Hall G

Fig. 4.12 Evaluation of daylight availability and brightness

are below the required threshold, supplementary artificial lighting is necessary to achieve the desired level.

The results of the daylighting levels can be seen in Fig. 4.12a–d and the luminance ratios in Fig. 4.11c, d, a Fig. 4.12e–f. More detailed results are provided in [1].

4.1.2 Case Study: Office Environment

Daylighting can significantly reduce energy consumption in offices. However, achieving this requires careful design to balance heat gains and losses, glare, and changes in daylight availability. While energy savings are crucial, it's also important to consider other related benefits, such as productivity and quality.

The preference for natural daylighting is attributed to the fact that working under natural light leads to less stress and discomfort compared to artificial lighting. The most desired indoor features in an office include good lighting, favorable thermal conditions, and ventilation. While good acoustics and adequate space are consistently high factors in various studies, they do not receive the highest ratings for satisfaction with the indoor environment. Dissatisfaction with daylighting in buildings is mainly caused by glare, especially from computer monitors. Therefore, special attention should be paid to glare during daylighting design.

The evaluation subject was an office in the Building Faculty in Košice. The office is located on the third floor, facing southeast with side lighting, as shown in Fig. 4.13a. Its dimensions are 3.4 m (width), 5.5 m (length), and 2.7 m (height). It is illuminated by side daylighting from a window measuring 1 200 × 1 800 mm. The sill height is 900 mm. The light transmittance was assumed to be 63%. Surface reflectances are: walls 70%, floor 20%, ceiling 70%. Daylighting analysis was conducted under overcast sky conditions, with illuminance values ranging from 10 500 to 25 000 lx. Internal illuminance was measured on a horizontal reference plane at a height of 0.75 m. Visual comfort under clear sky conditions was also evaluated.

In the room, simulations were conducted using the Velux Daylight Visualizer and RADIANCE software. Various viewpoints and positions of the work desk were evaluated due to excessive glare [2, 3]. Table 4.4 presents the CIE Daylight Factor (DF) values obtained from the measurements. This indicates that the measurements were used to calculate the daylight factor.

Brightness calculations are based on a model created through simulation and daylighting of the area. These brightness values are subsequently used in the glare calculation process. Comparing both simulation programs for evaluating brightness ratios reveals differences, although some values are comparable in certain areas (especially window brightness), as shown in Fig. 4.13b–e. Differences in brightness are attributed to the specified boundary conditions, which are more complex in the RADIANCE simulation program.

From Fig. 4.14 and beyond, it is evident that the brightness values are not within an acceptable range. In Figs. 4.15a, b and 4.16, potential glare sources are depicted. The issue lies in the orientation of the window to the south, leading to glare, necessitating the use of shading devices. Another option is to change the position of the desk, with the best position in this case being facing west.

Using RADIANCE simulation, the glare indices were evaluated as follows: for position A: DGP = 0.3; DGI = 25.3; UGR = 28.3. For position B DGP = 0.23; DGI = −14.7; UGR = −10.7. From the results, it is evident that the glare is unacceptable and intolerable.

From the results of the Guth VCP glare index, it is evident that for the southern view, people's satisfaction with visual comfort from 5 to 100% (Fig. 4.17a, b), for western view from 20 to 100% (Fig. 4.17c) and for eastern view from 10 to 90% (Fig. 4.17d).

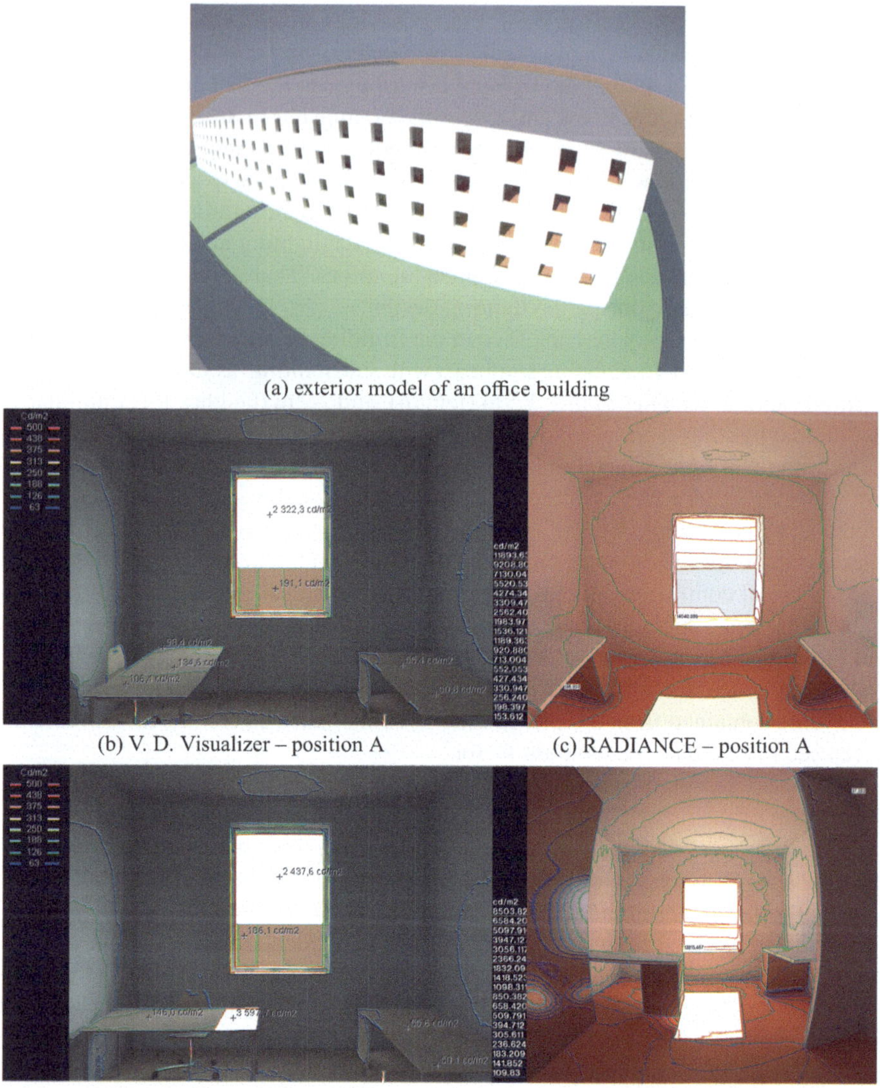

(a) exterior model of an office building

(b) V. D. Visualizer – position A

(c) RADIANCE – position A

(d) V. D. Visualizer – position B

(e) RADIANCE – position B

Fig. 4.13 Luminance values for a south–view

Table 4.4 Assessment parameters of daylighting—office

	Daylight factor		Uniformity of lighting
	D_{min}	D_{max}	U_0
Measured	0.52	4.59	0.32
Simulated	1.00	7.40	0.28

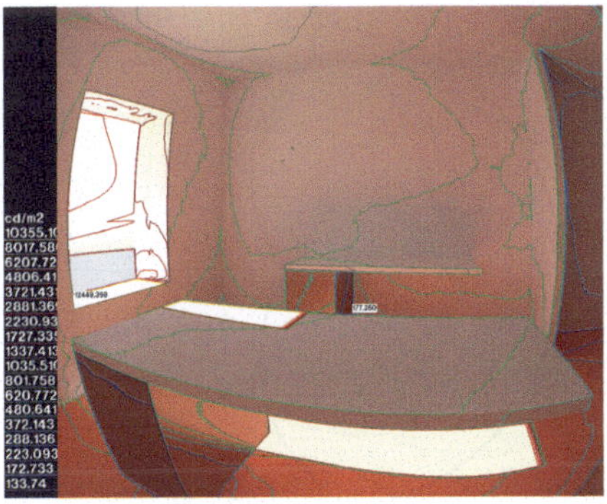
(a) western view

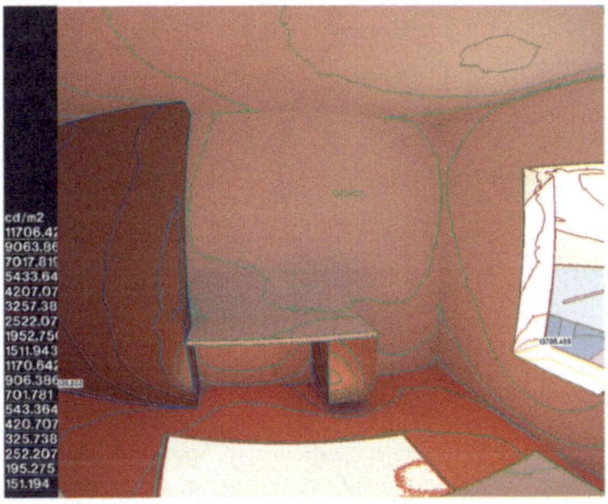
(b) eastern view

Fig. 4.14 Brightness values—position A

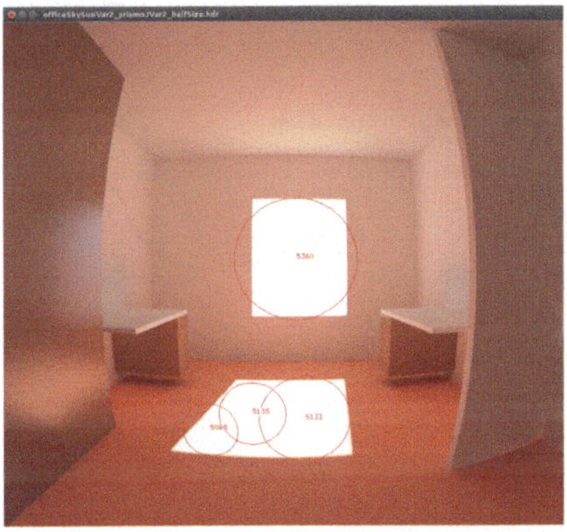
(a) southern view – position A

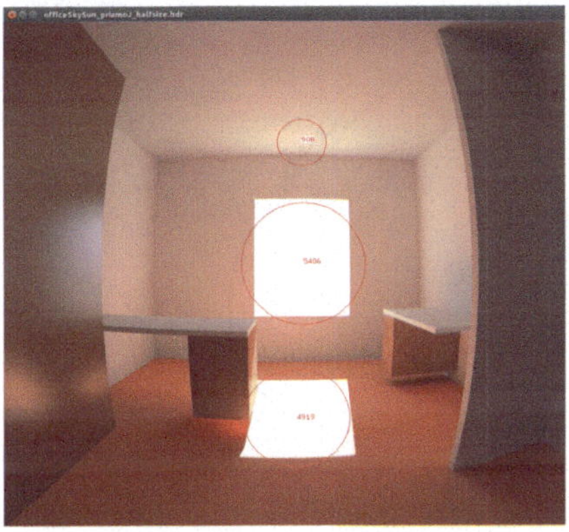
(b) southern view– position B

Fig. 4.15 Brightness values and glare sources

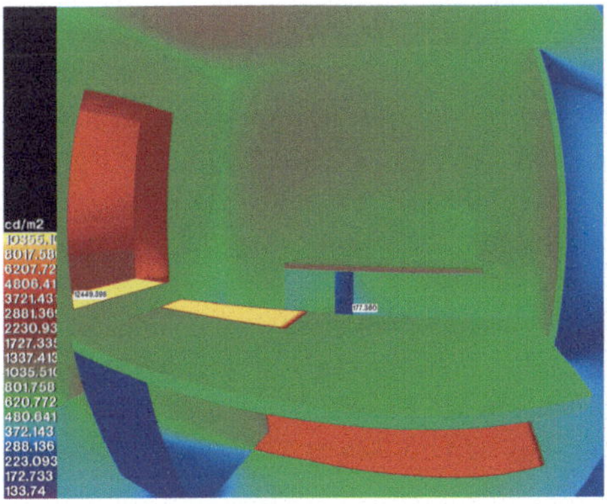
(a) brightness values — western view

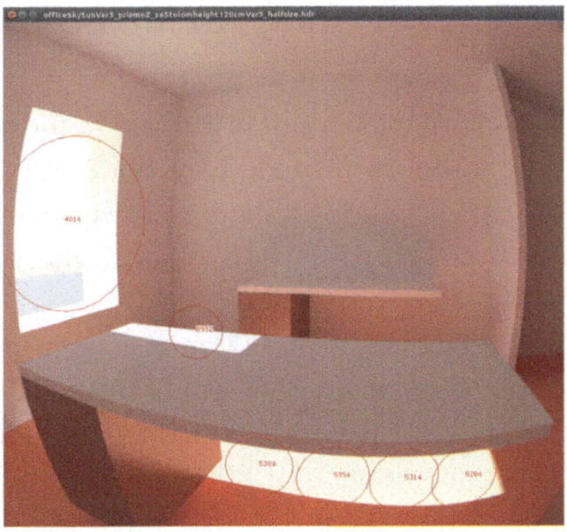
(b) glare sources — western view

Fig. 4.16 Evaluation of visual comfort

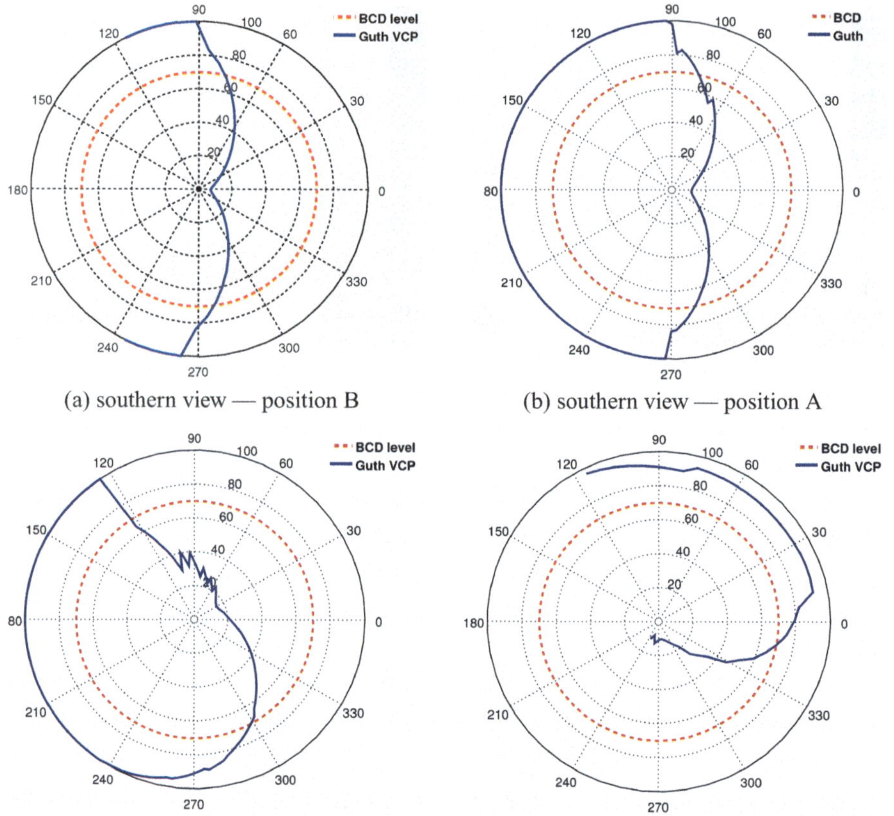

(a) southern view — position B
(b) southern view — position A
(c) western view
(d) eastern view

Fig. 4.17 Evaluation of visual comfort using the Guth VCP glare index

4.1.3 Case Study: School Classroom in the Attic

The school classroom is located in the attic spaces of the Building Faculty in Košice. Daylighting is provided by both side and top lighting systems. The side lighting consists of windows facing north, while the top lighting comprises skylights or roof windows (see, Fig. 4.18). The dimensions of the classroom are $10.77 \times 4.36 \times 3.03$ m.

The sill height is 900 mm. The windows measure 900×600 mm, and the roof windows measure 780×1180 mm. The window-to-wall ratio is between 0.05 and 0.12 without the roof windows included. The light characteristics considered are: light transmittance of 80%, reflectance of walls and ceiling at 70%, and reflectance of the floor at 45%.

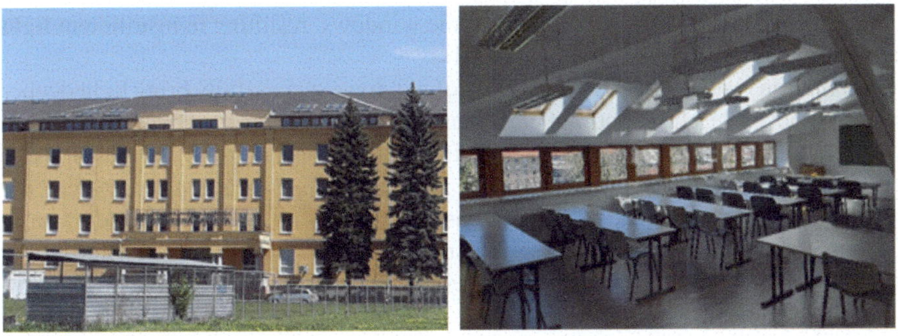

(a) brightness values — southern view (b) glare sources — southern view

Fig. 4.18 View of the building and the classroom

The height of the working plane is 0.8 m above the floor level. The light transmittance was considered to be 64%. The working hours were assumed to be from 8:00 AM to 5:00 PM.

For the purpose of planned reconstruction of the roof structure, daylight measurements were conducted in the room (refer to Fig. 4.19). The outdoor horizontal illuminance ranged from 5 000 to 7 000 lx on the selected day. Daylight factor was calculated based on the measured values. Subsequently, a room model was created and simulations were performed using the Velux Daylight Visualizer software for all selected alternatives (option without roof windows and three alternatives with different sizes and positions of roof windows).

The study conclusions revealed that the minimum value of the daylight factor calculated from the measured values does not meet the normative requirements, except for its average value, which is exceeded. The values obtained from simulations also do not fall within acceptable ranges. Internal illuminance was also evaluated, showing higher values near the side windows and directly under the roof windows. However,

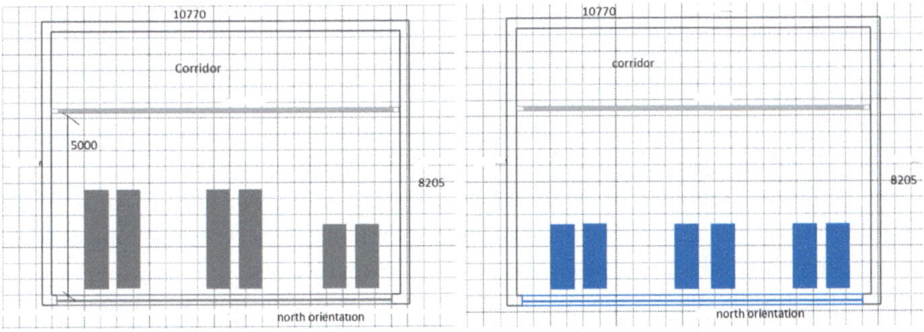

(a) basic model (b) optimal solution

Fig. 4.19 Window layout diagram

illuminance decreases further away from the windows, resulting in insufficient light levels for activities such as drawing (500 lx).

Simulation results demonstrated that roof windows successfully bring light deeper into the space. Therefore, optimal placement of roof windows was sought by varying their positions in the simulations. The simulations only adjusted the positions of the roof windows. The optimal solution appeared to be an alternative where the window sizes were reduced and kept consistent across all windows. In addition to inadequate daylight availability, the classroom faces overheating issues during the summer months. This problem was addressed in the study through two alternative solutions.

In the room, actual measurements of internal temperature were not conducted, hence the situation can only be considered a prediction. This issue could be part of further research. It is clear that sufficient light will be at the expense of excessive heat gain through roof windows and a roof covering with low thermal storage capacity. The highest considered temperatures in June and July were boundary conditions in the simulation evaluated for overheating.

From the conclusions, it follows that the maximum temperature inside the classroom is reached on June 20th for the variant representing the actual conditions 39.2 °C and for the variant without roof windows 31.6 °C. The difference between these variants is approximately 8 K. From the temperatures, one can also infer the effect of thermal storage capacity. The maximum outdoor temperature is reached on June 19th, but indoors, the maximum occurs on June 20th, influenced by the thermal storage behavior of the building envelope and its impact on the indoor environment of the attic space.

Since overheating occurs during the summer period, it is necessary to consider a shading strategy. A well-designed shading device will contribute to better daylight conditions. However, this alone will not reduce excessive overheating below acceptable internal temperature limits 27 °C. All characteristics of the roof construction and windows were used in both the current condition variant and the variant without roof windows in the overheating prediction. The forecast indicated that the internal temperature within the classroom environment exceeds regulatory limits in either case.

Taking into account very low air exchange, the internal temperature approaches 50 °C. If the air exchange is increased to 6 exchanges per hour, the internal air temperature will decrease by more than 10 °C. Further increases in air exchange rate are not significant because warm air enters the indoor space, which is heated by overheated surrounding structures.

The situation without roof windows is more favorable, but even then, it is not possible to meet the maximum prescribed temperature of 27 °C for indoor climate during summer. Cooling measures need to be considered.

From the results, the variant with shorter window dimensions was evaluated as the most optimal solution (see, Fig. 4.19b). The simulation results are shown in the Figs. 4.20 and 4.21.

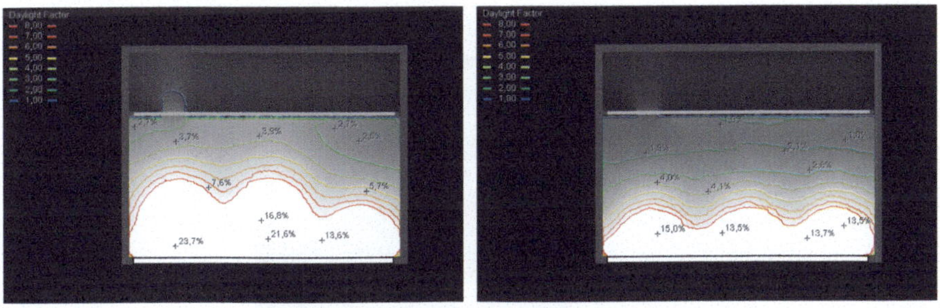

(a) basic model (b) optimal solution

Fig. 4.20 The availability profiles of daylighting

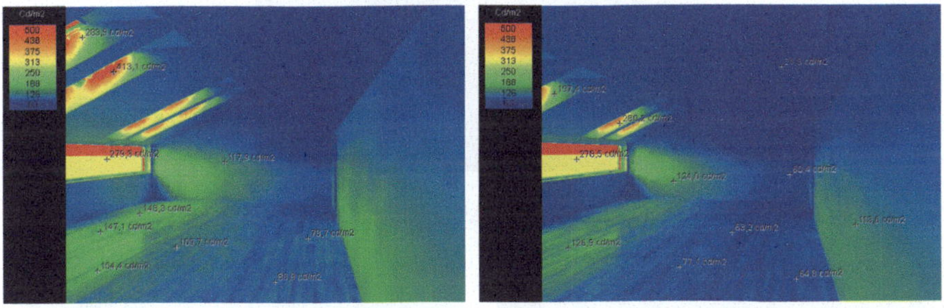

(a) basic model – luminance values (b) optimal solution – luminance values

Fig. 4.21 Simulation of visual comfort

4.1.4 Case Study of a Transparent Facade

Another studied object was an office located on the third floor of the Technicom building in Košice, depicted in Fig. 4.22. The building facade is designed as a double-skin transparent facade in the southern part of the building, incorporating adaptive elements that influence the indoor environmental conditions. The transparent facade extends along the entire length of the western front, partially mitigating adverse external environmental impacts.

The facade not only enables energy savings and noise insulation but also provides temporal variability to the architecture of the building through the adjustment of shading blinds' positions. Thanks to it, in the winter period, there are approximately 30% lower heat losses compared to a traditional transparent facade construction.

The double-skin facade on the southwest side serves to support natural ventilation by gravity. Its aerodynamic resistance can be regulated by intake and exhaust ventilation slots, as well as access grilles at different levels of each floor, which also serve for maintenance purposes.

It is designed as a shaft-type continuous along the entire height of the four floors. The air cavity serves to enhance the effect of gravity ventilation for air extraction,

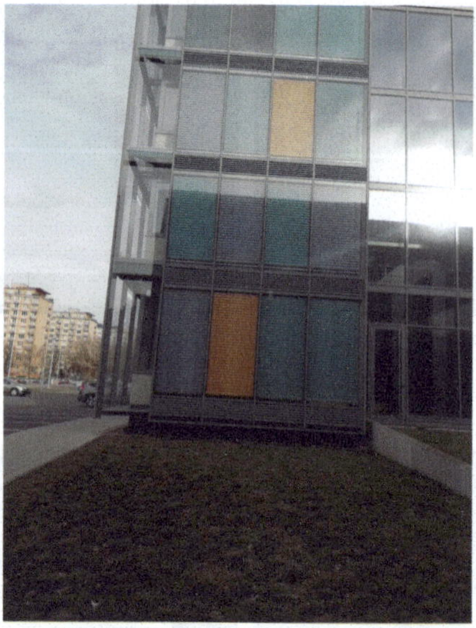

Fig. 4.22 External view from the south of the Technicom building

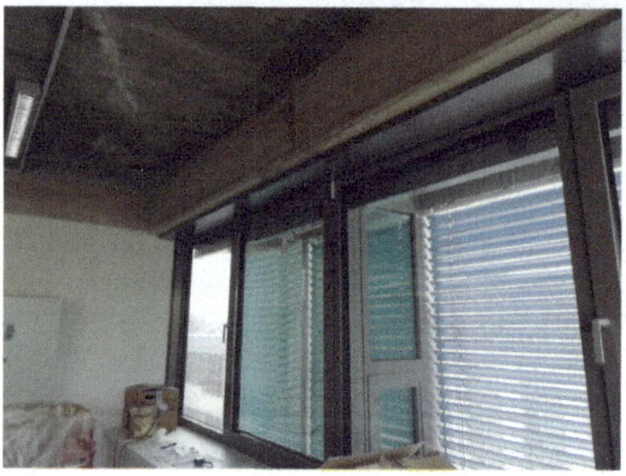

(a) east – south view

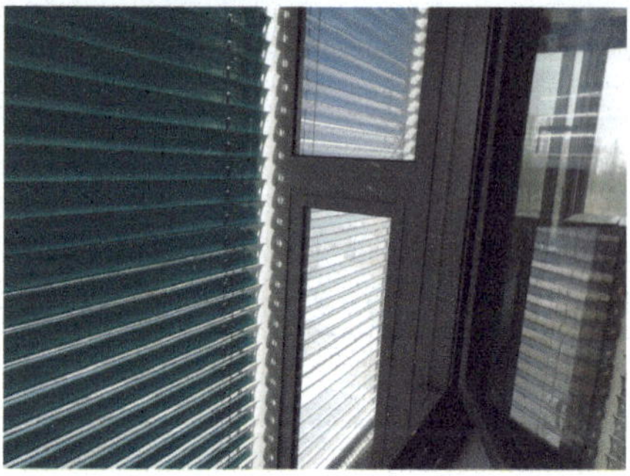

(b) view into the interstitial space

Fig. 4.23 Interior environment of the office

coupled with a shading system that will be used during the summer and transitional seasons based on solar radiation intensity. On the interior side, a double-glazed system with selective properties in both thermal and light spectra of solar radiation is used in the glazed sections.

The daylighting comfort was assessed in the corner room on the third floor of the building shown in Fig. 4.23. Daylight measurements were conducted in two different years, in January and March, with two measurements each at different positions of the blinds. Measurements were taken inside the room at nine points at a height of 0.85 m, corresponding to desk level (Fig. 4.24a, b). Measurement was conducted using KONICA MINOLTA CHROMA METER CL-200 with 2% accuracy and an

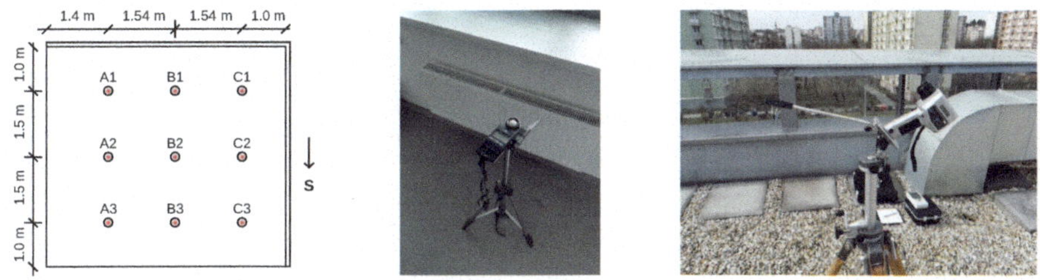

(a) measured points (b) indoor (c) outdoor

Fig. 4.24 Indoor and outdoor light measurements

illumination sensor ALMEMO FLA 623VL with serial number 15061543 and 5% accuracy.External horizontal illuminance was measured at the roof structure level at a height of 1.5 m (Fig. 4.24c) with the KONICA MINOLTA Luminance Meter LS-110, with an accuracy of 2%.

The measurement of the sky luminance distribution at the equal illuminance measurement location was conducted on both days at the beginning and end of the study. A comparison of the measured values with normalized values of sky luminance distribution under uniformly overcast sky conditions over a dark terrain is shown in Fig. 4.25. Outdoor illuminance levels ranged around 32 klx in March and from 4 700 lx do 8 000 lx in January.

From the graphs in Fig. 4.25, it can be observed that the sky luminance values at the measured elevation angles before and after the measurements did not always fall within the acceptable range. Specifically: in March, at the beginning of the measurements, the average sky luminance value at an elevation angle of 15° was 0.7 (standard range is 0.3–0.6), and at an elevation angle of 45° it was 0.81 (standard range is 0.7–0.85). At the end of the measurements, these values were 0.58 and 0.69, respectively. In January, at the beginning of the measurements, the average sky luminance value at an elevation angle of 15° was 0.58, and at an elevation angle of 45° it was 0.81. At the end of the measurements, these values were 0.50 and 0.71, respectively.

The resulting values of measured indoor illuminances at the control points for different blinds slope are provided in the Tables 4.5, 4.6 and 4.7. The calculated values of the daylight factor are listed in the Tables 4.8 and 4.9.

From the assessment of internal lighting measurements, it follows that at point A, the largest difference is at blind angles of 105° and 140°, and the smallest difference is at 32°. At point B, the largest difference is at angles of 32° and 105°, and the smallest difference is at 160°. At point C, there is no significant difference at all angles.

For further research on the behavior of the indoor climate, a model was developed and fine-tuned through precise measurements under specific conditions (which constitute the boundary conditions of the simulation process). This ensures the accuracy of predicting changes in the light microclimate within the considered space when

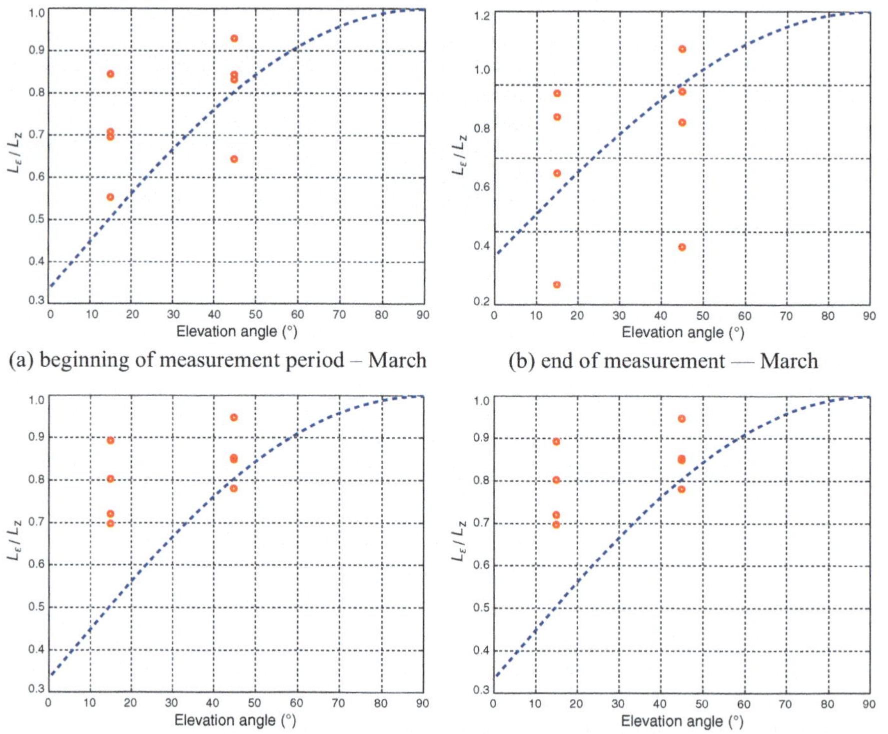

(a) beginning of measurement period – March (b) end of measurement — March

(c) beginning of measurement period – January (d) end of measurement — January

Fig. 4.25 Evalution of the distribution of sky brightness gradation over the days of measurement

Table 4.5 Daylight illumination for blinds slope 18° and 32°

	MAR		JAN		MAR		JAN		MAR		JAN	
	18°	32°	18°	32°	18°	32°	18°	32°	18°	32°	18°	32°
	A				B				C			
1	297	327	32	42	647	609	61	71	1733	1790	184	232
2	225	255	31	42	134	154	12	15	720	850	154	153
3	237	252	30	47	487	553	84	100	1623	1660	230	260

considering structural modifications (such as shading and others like various glass treatments, blinds, overhung canopies, ...) Evaluation of daylight availability and visual comfort was conducted using the simulation program RADIANCE.

In the calculations, the following coefficients were considered; see Table 4.10. There were no nearby objects that could shade the tested rooms. The transmittance factor of the glazing was measured 0.71.

Table 4.6 Daylight illumination for blinds slope 72° and 105°

	MAR		JAN		MAR		JAN		MAR		JAN	
	72°	105°	72°	105°	72°	105°	72°	105°	72°	105°	72°	105°
	A				B				C			
1	609	1982	82	163	984	2500	147	247	2140	3405	325	415
2	452	778	68	85	267	464	220	290	1050	1205	265	225
3	338	457	47	73	605	690	120	127	1780	1825	270	293

Table 4.7 Daylight illumination for blinds slope 140° and 160°

	MAR		JAN		MAR		JAN		MAR		JAN	
	140°	160°	140°	160°	140°	160°	140°	160°	140°	160°	140°	160°
	A				B				C			
1	2500	637	170	77	2400	900	240	118	3270	2050	350	280
2	654	300	70	49	370	190	132	116	970	815	260	238
3	385	267	61	44	645	570	115	105	1730	1680	308	263

Table 4.8 Daylight factor values (%)—March

Slope (°)	Daylight Factor DF_{min} (%)	Uniformity of illumination U_0 (–)
18	0.40	0.25
32	0.50	0.23
72	0.90	0.34
105	1.00	0.25
140	1.00	0.26
160	0.80	0.32

Table 4.9 Daylight factor values (%)—January

Slope (°)	Daylight Factor DF_{min} (%)	Uniformity of illumination U_0 (–)
18	0.20	0.11
32	0.30	0.16
72	1.10	0.41
105	1.10	0.35
140	0.80	0.31
160	0.50	0.27

Table 4.10 Values of the measured reflectance factors of the room surfaces (–)

Surface	Reflectance factor
Ceiling–concrete	0.50
Floor–gray	0.16
Wall white	0.50
Door white	0.17
Wardrobe gray	0.58
Door–gray	0.25
Partition between desks	0.60
Chair green	0.32
Chair blue	0.10
Monitor turned off	0.03
Desks	0.68
Window frame	0.10
Windowsill	0.87
Blinds–gray	0.39
Blinds–green	0.23
Blinds–blue	0.41

Table 4.11 Daylight factor for open blinds

	MEAS	SIM	MEAS	SIM	MEAS	SIM
	A		B		C	
1	3.0	3.3	4.1	3.1	6.3	6.3
2	1.4	1.4	0.7	1.3	2.0	4.6
3	0.9	1.0	1.5	1.6	3.3	3.9

In Table 4.11 are shown values of daylight factor from measured and simulated values without blinds. The results of the daylighting levels for simulated values can be seen in Fig. 4.26.

As can be seen from the values in the Table 4.11 and the Fig. 4.26, the minimum value of the daylight factor (0.6%) is met in almost all points for both the measured and simulated values. When comparing the measured and simulated values at individual points, it is clear that the largest difference between them is at point B2 and at a point C2, it is up to twice the simulated versus measured value. In other points it is comparable up to or from 6 to 32%. This indicates that the blinds have a significant impact on the level of daylight (quantity), especially the position (angle) of the blinds, which is evident in the additional images and tables for various blind angles.

The inclination of the blinds has a significant effect on the penetration of daylight into the interior, and the correct setting can optimize the amount of light that enters

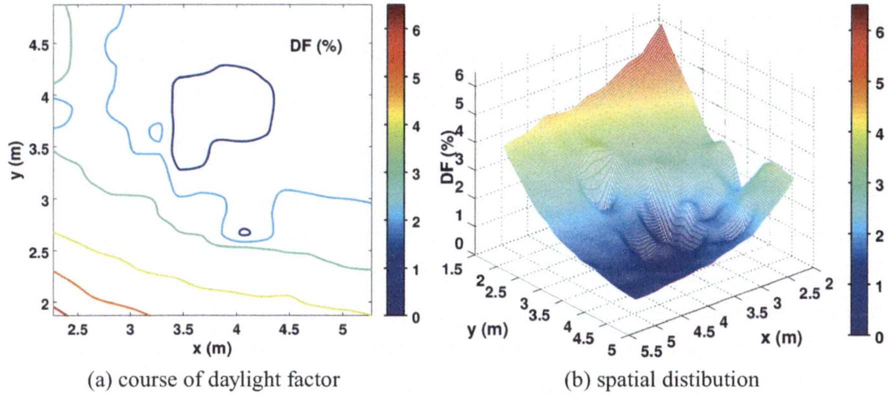

Fig. 4.26 Daylighting assessment without the influence of blinds—pulled up

the room, while minimizing glare and ensuring thermal regulation. The influence of the inclination of the blinds on the transmission of daylight:

- Horizontal position (0°)—maximum light transmission, as long as the slats of the blinds are horizontal, they allow maximum daylight to enter the room. This position is suitable when there is a need to illuminate the space with natural light as much as possible. The disadvantage of this setting is a higher risk of glare, especially when the sun is low in the sky (e.g. morning or evening).
- Moderate slope (15°—30°)—optimal balance, the moderate slope of the slats still allows sufficient light transmission, but at the same time limits the direct transmission of the sun's rays, which can reduce glare and heat gains. This tilt helps disperse light more evenly across the room, which can improve visual comfort and reduce the contrast between bright and dark areas.
- Steeper slope (45°—60°)—light control, the slope of the slats around 45° to 60° significantly limits the direct penetration of the sun's rays, which reduces glare and protects against overheating of the interior. However, the slope also reduces the total amount of daylight that reaches the room, which can be disadvantageous in spaces where light is desirable.
- Vertical position (90°)—minimum light transmission, when the slats of the blinds are almost vertical, the light transmission is minimal, which is suitable if the room needs to be darkened or if it is necessary to completely avoid glare. The vertical setting also provides the best protection against heat gain, which can be beneficial on hot days.

As for the optimal angle of the blinds, the most favorable angle of the blinds depends on the specific situation, including the position of the sun, the lighting requirements and the comfort of the user. In general, a slope between 15° and 30° provides a good balance between daylight transmission and minimal glare. This inclination allows sufficient light to enter the room and at the same time helps to direct the light so that there is no direct glare. For north-facing windows where

Table 4.12 Daylight Factor for blinds slope 18° and 32°

	MEAS		SIM		MEAS		SIM		MEAS		SIM	
	18°	32°	18°	32°	18°	32°	18°	32°	18°	32°	18°	32°
	A				B				C			
1	0.7	0.8	2.9	2.9	1.4	1.3	2.3	2.2	3.9	4.1	3.3	3.3
2	0.6	0.7	1.0	1.0	0.2	0.3	0.2	0.2	3.0	3.1	1.1	1.1
3	0.7	0.8	0.5	0.5	1.7	1.4	0.5	0.6	4.5	4.3	0.6	0.6

Table 4.13 Daylight factor for blinds slope 72° and 105°

	MEAS		SIM		MEAS		SIM		MEAS		SIM	
	72°	105°	72°	105°	72°	105°	72°	105°	72°	105°	72°	105°
	A				B				C			
1	1.3	2.5	3.0	3.1	2.3	3.7	2.4	2.9	4.9	6.3	3.5	5.7
2	1.1	1.3	1.1	1.2	3.2	4.3	0.4	0.9	4.0	3.4	1.3	3.8
3	0.8	1.1	0.6	0.8	2.0	1.9	0.8	1.2	4.5	4.9	0.9	3.1

Table 4.14 Daylight factor for blinds slope 140° and 160°

	MEAS		SIM		MEAS		SIM		MEAS		SIM	
	140°	160°	140°	160°	140°	160°	140°	160°	140°	160°	140°	160°
	A				B				C			
1	2.4	1.0	3.0	2.9	3.4	1.6	2.4	2.2	4.8	3.7	4.3	3.3
2	0.9	0.6	1.1	0.9	1.8	1.5	0.4	0.2	3.6	3.1	2.3	1.1
3	0.8	0.5	0.6	0.5	1.6	1.3	0.7	0.5	4.2	3.2	1.7	0.6

there is no direct sunlight, the slats can be opened more (horizontally) to maximize daylight. Conversely, for south-facing windows, where direct sunlight is intense, a slight inclination (e.g. 30°) may be more useful, which will help protect against glare and overheating. Morning and evening sun may require a greater inclination of the blinds to prevent glare, while at midday it may be appropriate to open the slats more.

The results of the daylighting levels for different blinds slope can be seen in Tables 4.12, 4.13, 4.14 and Figs. 4.27, 4.28 and 4.29.

All listed slopes and positions have values higher than 0.6%, except for points A3 (0.5%) and B2 (0.2%), which indicates that sufficient daylight is provided for these conditions. This is important for the comfort and efficiency of lighting in the interior, which is in accordance with the requirements for a healthy and comfortable environment.

Blinds pitches of 32°, 72° and 105° show higher DF values, which means that these pitches are more effective for maximizing daylighting in the space. This fact supports the idea that at higher slopes there is a better use of daylight. For all slopes, the differences between the measured and simulated values are almost comparable,

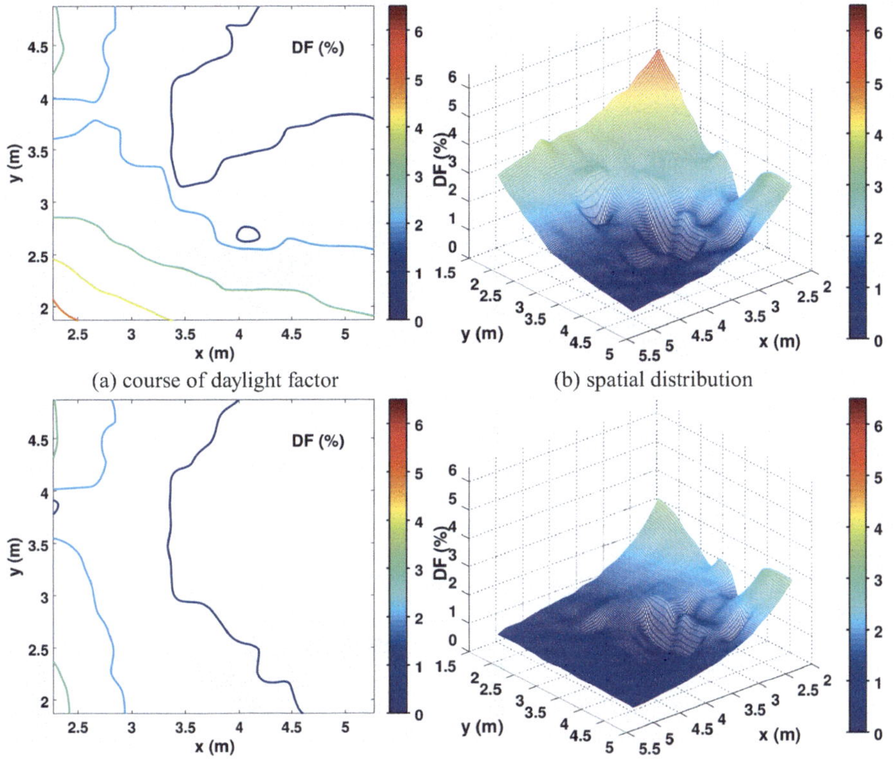

(a) course of daylight factor (b) spatial distribution

(c) course of daylight factor (d) spatial distribution

Fig. 4.27 Evaluation of daylight illuminance for blind slope 18° (**a**, **b**) and 32° (**c**, **d**)

ranging from 0.1 to 2.8%. The exceptions are the DF values at points C3 at slopes of 18°, 32° and 72° (about 3.5%) and at point B2 (3.4%) at a slope of 105°, which indicates that the simulations for angles of 140° and 160° are more accurate. At an inclination of 72°, the values are significantly higher, which indicates better lighting.

However, the differences between simulated and measured values start to become more pronounced, which may indicate the need for simulation calibration or better optimization for this slope. The slope reaches the highest daylight values, but the differences between the simulated and measured values are the largest, which may indicate the nonlinear behavior of the simulation model at higher slopes. The 105° slope achieves the highest daylight values, but the differences between the simulated and measured values are the largest, which may indicate the nonlinear behavior of the simulation model at higher slopes. Inclinations around 72° and 105° could be preferred in areas where it is necessary to ensure a high level of daylight, for example in office spaces or in residential areas with high requirements for natural light.

Based on the values mentioned, it is possible to recommend slopes between 32° and 105° as optimal for ensuring sufficient daylight, taking into account the used sim-

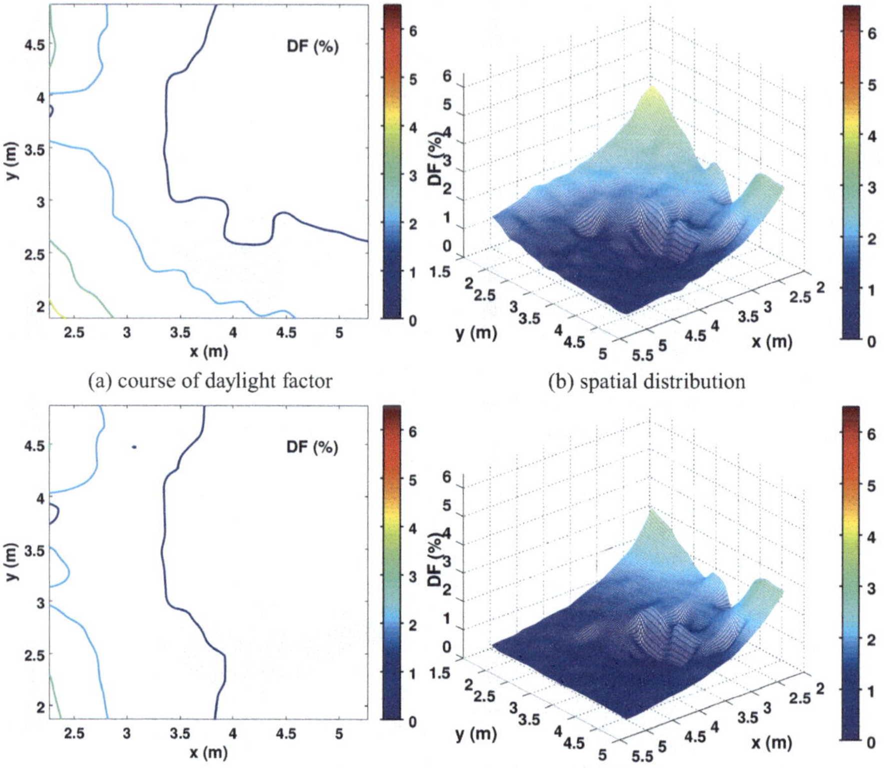

(a) course of daylight factor (b) spatial distribution

(c) course of daylight factor (d) spatial distribution

Fig. 4.28 Evaluation of daylight illuminance for blind slope 72° (**a**, **b**) and 105° (**c**, **d**)

ulations and measurements. Specific environmental conditions and lighting requirements should be considered for precise planning, with slopes between 72° and 105° providing the highest levels of illumination that might be preferred in environments requiring intense daylight. Overall, the 72° slope represents a good compromise between daylight intensity and simulation accuracy, making it potentially optimal for most conditions. If an even higher light intensity is required, a slope of 105° may be considered, but with the understanding that the accuracy of the simulation may be lower.

Differences between measured and simulated values can arise for several reasons. Simulation models often use simplified assumptions that can lead to deviations from actual conditions. A model may not accurately capture all physical processes that occur in the real world. The simulation software used may have limitations in accuracy, especially when it comes to complex environments or variable lighting conditions. The measurements themselves could be affected by changes in the surrounding environment, such as moving objects, changes in lighting, or atmospheric conditions that are not included in the simulation. Every measurement may con-

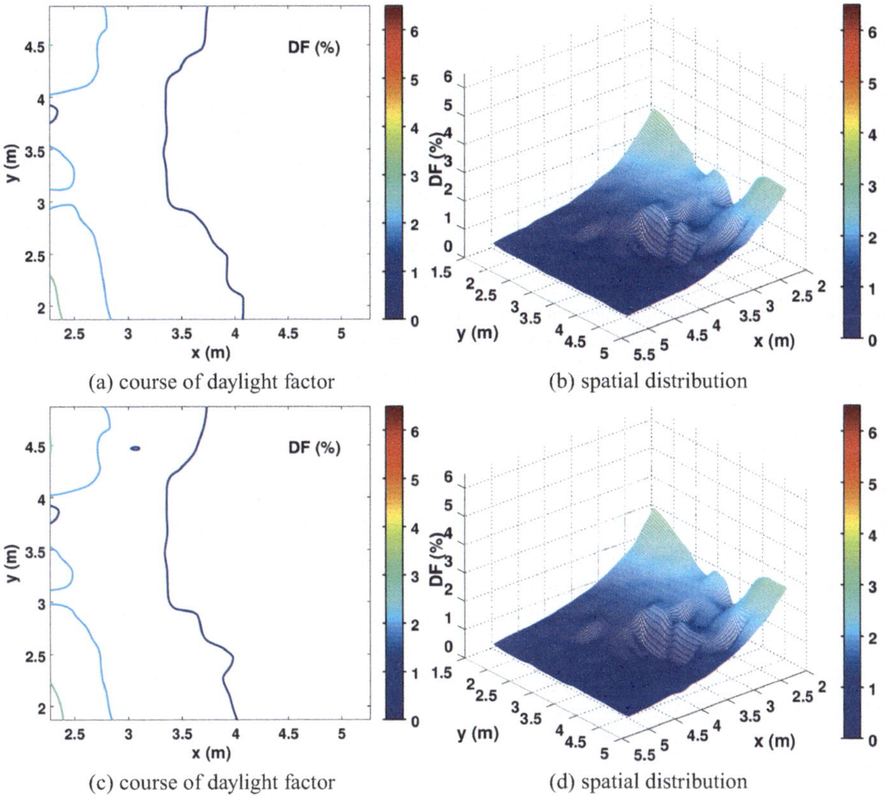

(a) course of daylight factor
(b) spatial distribution
(c) course of daylight factor
(d) spatial distribution

Fig. 4.29 Evaluation of daylight illuminance for blind slope 140° (**a**, **b**) and 160° (**c**, **d**)

tain a certain degree of error or inaccuracy due to the technology used, instrument calibration, or human error. If the input parameters to the simulation (e.g. material properties, space geometry) are inaccurate or incomplete, the simulation results can differ significantly from real measurements. If the simulation uses simplified models of geometry or material properties, this can lead to incorrect results. If the points where the measurements were taken are slightly different from the places where the simulation predicts the measurement, this can cause differences in the daylight value. The simulation may not include all the specific conditions that were present during the measurement, for example the exact angle of the sun's rays at a given time or the exact orientation of the surface. These factors can cause the simulated values to be either higher or lower than the measured values, leading to differences between them. A detailed analysis of all these factors would be necessary for an accurate comparison.

References

1. D. Katunský, E. Dolníková, B. Dolník, Sustainability **10**(3), 655 (2018). DOI https://doi.org/10.3390/su10030655. URL http://dx.doi.org/10.3390/su10030655
2. E. Dolnikova, Electronic Journal of the Faculty of Civil Engineering Osijek-e-GFOS **10**(18), 82 (2019). DOI https://doi.org/10.13167/2019.18.8
3. E. Dolníková, D. Katunský, (19th International Multidisciplinary Scientific GeoConference SGEM 2019, 51 Alexander Malinov blvd, Sofia, 1712, Bulgaria, 2019), vol. 19, pp. 597–604. DOI https://doi.org/10.5593/sgem2019/6.2/S27.076

Conclusion

The proposed chapter focuses on the assessment and prediction of lighting comfort in buildings. It highlights the phenomenon of daylight and its effects on indoor environments. Along with heating, lighting is a primary factor in buildings that influences user comfort, productivity, and health. Natural lighting is one of the most important design elements in indoor environments.

While electric lighting systems can provide useful illumination in the absence of daylight, allowing workers to extend their working hours, they also create various physical, physiological, and psychological issues.

Daylight is crucial for visual comfort as it effectively stimulates both the human visual and circadian systems. Despite these benefits, daylight cannot always provide the best visual performance; it may cause visual discomfort through glare and diffusion. Additionally, daylight can diminish the performance of the visual system by masking reflections or shadows in the workplace. The effectiveness of daylight for visual performance depends on its quality, which is similar to that of electric lighting. Therefore, both positive and negative aspects of daylighting exist.

The chapter is systematically divided into four sections. It begins with a theoretical analysis of the current state and a synthesis of knowledge in the area of computational methods and lighting assessment. Next, it offers a classification of simulation methods and the simulation program RADIANCE.

Today, advanced computing power and modeling techniques are used in all scientific fields, including lighting technology. Simulation programs are successfully employed to study lighting comfort in various buildings. These programs can model the daylighting conditions of indoor spaces.

The goal of modeling is to create a model that closely represents real operational conditions. For this, it is necessary to use actual measured data to ensure the model is created with minimal deviation.

© The Editor(s) (if applicable) and The Author(s), under exclusive license to Springer Nature Switzerland AG 2025

E. Dolníková, *Simulation Tools for Predicting Daylighting Conditions in Buildings*, Digital Innovations in Architecture, Engineering and Construction,
https://doi.org/10.1007/978-3-031-78330-2

The last two sections focus on two interrelated areas. The first area involves measuring daylighting in real buildings. The second area involves modeling these buildings using computational techniques. Based on measurements, the availability of daylight and visual comfort were evaluated.

Various types of industrial buildings were assessed, including large, low-ceiling spaces (textile factories) and smaller, high-ceiling spaces (engineering operations). In selected areas of a textile factory located in eastern Slovakia, daylighting needed to be evaluated. After conducting preliminary calculations, in situ measurements, modeling, and calculations using simulation programs, the following findings were presented.

A common parameter for daylighting in many countries is the daylight factor. Measuring daylight levels is quite complicated because outdoor lighting levels can change throughout the day. Finding a suitable day with acceptable conditions is difficult. Even when the criteria are met and outdoor lighting is within the permissible range, it can change during measurement, especially when many control points are involved. The case study demonstrates how to manage this task. In regular residential buildings, it is easier since the space is relatively small.

However, in industrial buildings, the interiors are large. Manufacturing halls are sometimes higher, and top daylighting (skylights) can only be used in the floor under the roof. In lower floors, structures can be modified or light pipes used. If only windows are used (as in this case), the daylight factor decreases significantly.

In this case, the hall was an industrial operation measuring 54×54 m with a height of about 5 m. This type of "hall" can be compared to large spaces in other buildings (not just industrial ones). Such spaces are also typical for administrative and other buildings, where free arrangements without built-in partitions are possible. In this study, a simple relationship was derived to predict daylight levels based on window area. This relationship can be used to predict the size of glazed sections in the exterior wall for given areas. Since windows alone are insufficient for daylighting, light needs to be brought in by other means over greater distances.

As we know, daylight is physiologically and psychologically essential. People inside must feel a connection between the enclosed interior and the exterior. This is sometimes just a visual idea that cannot be mathematically described. Therefore, the need for daylight in the conditions we described is essential when the daylight factor is low.

Calculations conducted and compared with measurements in actual industrial operations are presented in this case study of large spaces. It should be emphasized that the results differ when considering normative boundary conditions, as with simulation program calculations. Differences also exist when considering furnished versus unfurnished interiors. Any situation of preliminary determination of furnishings and built-in elements in the interior will introduce additional conditions, and only the calculated results will be indicative. They approximate the values we can expect in an actual interior.

When evaluating the sustainability of buildings based on computational programs, daylight becomes just one of the indicators of sustainability. The need for daylight is also essential for sustainability, and its emphasis should be highlighted in evaluation programs. Finding a compromise between ecology, the environment, economics, and energy requirements is not easy. However, this is a task for all of us in the future, as moving closer to sustainable development is our goal.

The following results were achieved in the engineering operation area. According to the literature used, the allowable deviation of the relative luminance of the sky at a given altitude from the zenith luminance for the given case for 15% was 0.3–0.6, and 45% in the range of 0.7–0.85. Interior lighting measurements are almost within acceptable ranges and represent daylight under adverse cloudy conditions. According to the assumed conditions, daylighting is considered satisfactory due to the limited measurement accuracy if the measurement results do not differ by more than 15% from the required values.

It is clear that visual comfort is reduced for both types of skylights. The luminance ratios of the individual visual tasks assessed are not within the maximum luminance limits. However, there is also a risk of glare caused by both roof windows.

During daylight measurement, it is challenging to meet standard conditions under cloud cover. The evaluation of measured data without corrections for comparable daylight conditions is questionable. A simple method of normalizing measured data to theoretical values representing cloudy sky conditions was proposed and tested. This allows for comparison of the data obtained from halls and their evaluation.

The halls showed similar average daylight levels. The hall with smaller lighting openings but an arched skylight shape was better lit by daylight. This is indicated by the results of the measured minimum and maximum daylight factors, as well as a higher level of lighting uniformity.

For more precise observation of details on the working plane, it will be necessary to supplement daylight with artificial lighting in this space.

Optimizing energy consumption in industrial buildings, along with internal environmental comfort for occupants, is a crucial issue when considering new standards and regulations as well as energy costs.

In a further study, the effectiveness and adequacy of integrated lighting in large industrial buildings were examined and analyzed to improve the quality of the indoor environment and meet relevant standards. Emphasis was placed on using daylight to enhance visual comfort in the studied industrial hall during the day while also saving energy. It was found that daylight in the studied knitting hall of the textile factory did not meet the required conditions throughout the workspaces.

At points and locations where the values are below the required illumination level, artificial lighting, such as electric bulbs, needs to be provided to supplement daylight and achieve the desired level. The minimum required daylight factor values for different types of work activities were determined based on the national standardization method ČSN 360011.

If the building designer does not have an appropriate calculation method, it would be difficult to determine the required lighting area in a specific production plant. The studied production area was very large, so it was necessary to calculate daylight factor values at many control points distributed according to the ČSN 730580-1987 and ČSN 730580-2000 standards. The lighting system design would meet the required daylight factor at specific distances from windows during the day. The measured daylight factor values were found to be very close to the skylight component of the overall lighting. The skylight component was observed to average 30% above the measured daylight factor values.

In industrial building design, daylight is not emphasized, and its contribution is usually small, but it is a crucial element for the psychology and physiology of workers. Workers need to feel connected to the outside environment; otherwise, their productivity decreases. It is advisable to reference other studies focusing on different window orientations and other influencing parameters, as well as on the visual comfort of workers and the effects of natural lighting on their productivity in the future.

In school buildings, a classroom located unconventionally in the attic was evaluated. Students spend more time at school than anywhere else. About 70% of this time is spent in classrooms. Daylight significantly impacts students' learning outcomes. Proper use of daylight can improve the indoor environment and reduce energy consumption.

Windows allow natural light to enter the interior. In the vertical plane of the building envelope, in the windowsill, they allow visual connection with the outside environment. Energy savings from daylighting are achieved by reducing the use of electric lighting. Daylight also increases passenger satisfaction and comfort. Windows are a complex part of building design and provide significant benefits to school buildings.

The study aimed to evaluate the availability of daylight in the attic, which serves as a teaching space. The study compares three alternatives for skylights. Roof windows are certainly necessary to provide more light to remote areas. Interestingly, moving the roof window strip achieves maximum uniformity and the required levels even at distances from the perimeter walls.

The study used simulation programs for model window situations. Based on the simulations of the daylight factor and predictions of overheating, the results were integrated closer to an optimal state. The results of this study provide recommendations for the appropriate design of roof windows in attics and their use for similar studies. The entire process was carried out in collaboration with lighting and thermal conditions.

The measurement results indicate that the interior lighting levels in the basic variant are below the standard. Informative measurements of indoor temperature during the summer show that a high indoor air temperature was reached, causing discomfort. Daylight was insufficient to ensure good lighting, and the illumination was not evenly distributed. Despite increased ventilation, the temperature remained excessive. The analysis showed that there are many alternatives regarding the placement of openings to improve daylighting in the classroom.

It was found that a combination of façade and roof windows (combined lighting) yielded better results, and that vertical windows are not very important for calculating temperature increase in the summer. Attic spaces are not commonly used in the teaching process. Only a few classrooms are located in attics. The study shows that the school expanded its teaching space by raising the roof and creating new classrooms.

There is an excessive amount of light in the classrooms; the problem is overheating in the summer. The prediction of the indoor climate during the summer period showed that the proposed design solution could not achieve the prescribed comfort. Shading elements help reduce overheating and lower the indoor temperature but not below the prescribed maximum indoor air temperature. The design solution does not solve the problem; cooling needs to be designed.

For the assessed period (June, July), cooling is proposed to ensure a sufficient reduction in indoor temperature during the summer to the desired level. The graphical dependency derived in the study serves to predict similar spaces in connection with achieving the required daylight factor and estimating the attainment of indoor temperature in the summer.

Offices in administrative buildings are primarily evaluated in terms of good visual comfort. Visual comfort and productivity are closely related. This is not surprising, as the better the visual comfort, the higher the productivity of workers. Besides visual comfort, productivity is affected by factors like size, contrast, source brightness, and the time it takes. The greater the impacts, the higher the productivity.

Visual comfort parameters were evaluated by modeling the office space using Velux Daylight Visualizer and Radiance software. The first software allows for the calculation of brightness values, daylight factor. The second provides data on DGP, DGI, UGR, VCP, and others.

A DGP value that is unacceptably glaring is higher than 45, a DGI value is higher than 31, and a UGR value is higher than 28. The higher the UGR value, the greater the damage to human eyes. The calculated UGR value exceeds the threshold by up to 70% at a given point (the limit for offices is 19).

In some cases, there may be disturbing reflections. A window located on the southern orientation leads to critical issues with glare and necessitates the use of shading devices. Another solution would be to change the position of the desk to minimize the level of glare.

The contribution of this chapter lies in the processing and deepening of current knowledge in the examined area, their classification, analysis, and generalization for the development of simulation program theory in the field of daylight.

For practical application, the work contributes by processing and explaining the given issue in case studies, as well as by outlining goals for further development and perspectives in the given area.

In conclusion, it can be stated that the chapter outlines further development in the field of designing daylight in building interiors.

The manufacturer's authorised representative in the EU is Springer Nature Customer Service Centre GmbH, Europaplatz 3, 69115 Heidelberg, Germany. If you have any concerns regarding our products, please contact ProductSafety@springernature.com

Printed and bound by CPI Group (UK) Ltd, Croydon, CR0 4YY
01/10/2026
02225983-0009